New Deep Territories

Oceans in Depth

A SERIES EDITED BY KATHARINE ANDERSON
AND HELEN M. ROZWADOWSKI

∴

New Deep Territories

∴

A STORY OF FRANCE'S EXPLORATION
OF THE SEAFLOOR

Beatriz Martinez-Rius

THE UNIVERSITY OF CHICAGO PRESS
CHICAGO AND LONDON

PUBLICATION OF THIS BOOK HAS BEEN AIDED
BY A GRANT FROM THE BEVINGTON FUND.

The University of Chicago Press, Chicago 60637
The University of Chicago Press, Ltd., London
© 2026 by The University of Chicago
Published 2026
Printed in the United States of America

35 34 33 32 31 30 29 28 27 26   1 2 3 4 5

ISBN-13: 978-0-226-84637-8 (cloth)
ISBN-13: 978-0-226-84639-2 (paper)
ISBN-13: 978-0-226-84638-5 (ebook)
DOI: https://doi.org/10.7208/chicago/9780226846385.001.0001

Library of Congress Cataloging-in-Publication Data

Names: Martinez-Rius, Beatriz, author
Title: New deep territories : a story of France's exploration of the
    seafloor / Beatriz Martinez-Rius.
Other titles: Oceans in depth
Description: Chicago : The University of Chicago Press, 2026. | Series:
    Oceans in depth | Includes bibliographical references and index.
Identifiers: LCCN 2025033001 | ISBN 9780226846378 cloth | ISBN
    9780226846392 paperback | ISBN 9780226846385 ebook
Subjects: LCSH: Ocean bottom—Discovery and exploration | Research—
    France | Marine resources
Classification: LCC GC87 .M36 2026
LC record available at https://lccn.loc.gov/2025033001

♾ This paper meets the requirements of ANSI/NISO Z39.48-1992
(Permanence of Paper)

To my husband,
Javi

Contents

Foreword: Oceans in Depth

In the mid-twentieth century, French exploration of the ocean went to the bottom—to the seafloor with the marine geosciences. Drawing on a wealth of archival resources of institutions for science and policy, the oil industry, and individual scientists, Beatriz Martinez-Rius tells the story of a distinctive French vision of oceanic territory. She shows how ocean exploration depended on a compelling mixture of ideas about both the past and the anticipated future. These included both the historical identity of France as a maritime power and its place in the post-1945 international order. In an age of decolonization, such exploration developed ideas and practices for a new kind of territory under the surface of the water. With the Mediterranean as a model, the interests of French scientists, industry leaders, and politicians converged on the geological origins of sea basins, making ocean sciences—and potential ocean resources—an anchor for France's geopolitical identity.

The book keeps the volumetric oceans, real and imagined, at its center. Oceanic territory and resources motivated French projects to extract knowledge about the seafloor and seabed beneath it. These activities opened a new chapter in the human relationship with this remote and challenging environment. Against the backdrop of US dominance of ocean sciences and the offshore oil industry, Martinez-Rius shows how France became an unexpectedly important player in both the development of ocean technologies and deep-sea oil and mining ambitions. Through science and diplomacy, French actors built strategic collaborations with the US and Japan, including participation in the Deep Sea Drilling Project (1968–83).

The French perspective sharpens our view of the international dimensions of oceans. In the same era, debates accelerated worldwide about oceans as international space or national territory, leading to the United Nations Law of the Sea Convention in 1982. Charismatic figures like Jacques Cousteau spurred the public imagination in France and beyond, turning underwater environments into places that could be filled with human activity. As she traces the exploration of deeper and deeper waters in the Atlantic,

Pacific, and Mediterranean, with scientific projects in ever closer collaboration with industry, Martinez-Rius builds a compelling picture of international networks motivated by scientific curiosity, industry patronage, and geopolitics.

This accessible work brings complicated subjects—marine geosciences, deep-sea technologies, and the distinctive public-private setting of scientific and economic policy in the French state—to life. More importantly, Martinez-Rius shows how the connections between all these subjects mattered to France and its citizens in the twentieth century. Her work gives a compelling account of why these technoscientific networks continue to demand our attention today.

Katharine Anderson
Helen M. Rozwadowski

Acronyms

AFERNOD	French Association for the Study and Research on Oceanic Nodules
ANZIC	Australia and New Zealand International Scientific Drilling Consortium
BRGM	Geological and Mining Research Office
CEA	French Atomic Energy Commission
CEPM	Committee of Petroleum and Marine Studies
CFP	French Petroleum Company
CIESM	International Commission for the Scientific Exploration of the Mediterranean
CNES	French National Center of Space Studies
CNEXO	Centre National pour l'Exploitation des Océans (National Center for the Exploitation of the Oceans)
COB	Oceanological Center of Brittany
COMEXO	Committee "Exploitation of the Oceans"
DGRST	General Delegation for Scientific and Technical Research
DSDP	Deep Sea Drilling Project
ECORD	European Consortium for Ocean Research Drilling
Elf-ERAP	Essence Lubrifiants France–Entreprise de Recherches et d'Activités Pétrolières
ERAP	Entreprise de Recherches et d'Activités Pétrolières
IACOMS	International Advisory Committee on Marine Sciences
ICES	International Council for the Exploration of the Sea
IFP	French Institute of Petroleum
IIOE	International Indian Ocean Expedition
IOC	Intergovernmental Oceanographic Commission
IODP-1	Integrated Ocean Drilling Program
IODP-2	International Ocean Discovery Program
IODP-3	International Ocean Drilling Programme, called "IODP cubed"

NERC   British Natural Environment Research Council
OPEC   Organization of the Petroleum Exporting Countries
ORC PEPE   NATO's Subcommittee on Oceanographic Research
SCOR   Special Committee on Oceanographic Research
SNPA   National Society of Aquitanian Petroleum
UNCLOS I   First United Nations Convention on the Law of the Sea
VNIRO   Russian Federal Research Institute of Fisheries and
                Oceanography

A World's View from
the Seafloor

From an early age we learn to recognize the contours of the world's countries. In the maps hanging on the walls of our classrooms and printed in our textbooks, the boundaries that crisscross continents demarcate regions subjected to human conflicts and alliances, mutable throughout history. Yet beyond the line representing the coastline, the oceans expand as blank surfaces detached from human activities, timeless and oblivious to the changes and events happening on the mainland. Not many of us would be able to draw accurately the maritime frontiers of coastal countries—not even that of our own. And that ocean space, its volumetric mass of seawater and its three-dimensional seafloor, remains stubbornly outside our worldview.

Yet beyond the blue surface of seas and oceans, and beneath their vast waters, lies a territory as important for human societies as exposed lands and the airspace above them: the seafloor. Even though it is uninhabitable and remote, our daily life is inevitably linked to the seafloor and its resources, a significance that transforms it into a three-dimensional space of paramount political, military, and economic importance. Ninety-nine percent of global telecommunications, including the internet, travel through underwater cables laid over the seabed (if you are reading this online, data coded for this text has probably just arrived from across an ocean).[1] One-third of the world's hydrocarbon supplies are extracted from the seafloor, especially over shallow continental shelves. Saudi Arabia in the Persian Gulf, Brazil across the southern Atlantic Ocean, and Mexico over its Gulf top the list of offshore oil-producing countries, followed by Norway in the North Sea and the US in the northern Gulf of Mexico. In the years to come, the apparent quietness of the greater depths might also be disrupted by a renewed international frenzy to secure strategic minerals, key to developing modern electronic systems. Deep-sea mining has become a hot topic among coastal governments, who seek to control potential deposits of manganese, nickel, or cobalt thousands of meters deep. In international forums like the United Nations (UN), the legislation to regulate deep-sea mining in areas beyond national jurisdiction is undergoing press-ing negotiations.

Why does the seafloor occupy such a minor position in the popular worl-dview and territorial understanding? The answer may lie in our incapacity to directly perceive it, the inaccessibility of interacting with the seafloor without extraordinary technological mediation. We cannot inhabit or travel across it. Only through advanced, exclusive, and costly technologies like submersibles, drilling vessels, or marine geophysical devices can the seafloor be known, charted, studied, and controlled. Water and darkness complicate the crafting of a public imaginary for the terrain under the sea.[2] When imagining the seafloor, we tend to recall images picturing narrow undersea areas illuminated by powerful beams of light, surrounded by the most acute darkness. Fantastic scenarios depicted in science fiction books and films have worked out the rest of this alien landscape.[3] In the popular imaginary, we possess a fuller image of the surface of the moon or Mars than we have for the seafloor. We do not even own a word to name the sub-seafloor (its underground), from which natural resources like oil and gas are extracted. For this reason, in this book, I use "seabed" to refer to the seafloor's surface, and "seafloor" to indicate surface *and* underground. The seafloor's neglect in the popular worldview has dramatic consequences for our understanding of the past and the present, as it obscures the role that the quest to control this space and its hidden natural resources has played, and will play, in international relations—both frictions and alliances—and national strategies of economic development.

This book is a call to situate the seafloor both in our worldview and equally in the history of the twentieth century. Its central idea is that, re-gardless of its remote and uninhabitable nature, the seafloor *is* a territory: an area shaped by political, economic, and strategic processes; tamed, known, and controlled via technoscientific practices, and explored and exploited through human labor and complex infrastructures.[4] Like any terrestrial territory, it has a history of its own, one that accounts for the processes that unfolded to legislate and rule over it, embraces the needs and motivations that triggered its exploration, and acknowledges the pub-lic imaginaries that acted as a fuel for expanding knowledge about it. Such a history must focus on the technoscientific practices deployed for sur-veying, controlling, and eventually utilizing the seafloor. It must also trace the emergence of new expertise and cooperative relationships between different communities to study it. And, finally, it must chart the emergence of this new territory, full of promises of riches and expanded national con-trol, which has influenced international relations and geopolitical power games. That period is the focus of this book. Humans have explored new territories since prehistoric times, chasing more benign climates and improved living conditions, fleeing from natural catastrophes or following herds of prey, looking for new sources of food and natural resources, or contesting established power relations. At the end of the Second World War, the exploration of the seafloor was propelled by equivalent motivations, powered by new economic imaginaries that flourished in the wake of decolonization.[5]

Early in the 1960s, the independence of regions under colonial rule and the crumbling of colonial empires spurred the emergence of global anxieties about overpopulation, resource erosion, and environmental degradation on land. Concerns and priorities derived from Cold War dynamics and the East-West fracture added to this backdrop, driving the attention of coastal governments and private industries toward the oceans as territories to be scouted, explored, mapped, controlled, and connected to mainland economies. Innovations in marine technologies, frequently deriving from underwater military devices improved during the two world wars, progressively opened new spaces of exploitation. In 1969, the legal scholar Edward Miles emphasized that, until the previous decade, uses of the oceans were essentially restricted to its surface, but that new technologies were now opening the threshold of the "third dimension": the depths and the seafloor.[6] The potential but barely explored resources from the seafloor came to be deemed economic pillars for future generations, as sources to secure energy and material supplies external to the land territory of nations. Scientists, engineers, and political leaders described a cornucopia of valuable materials hiding in and below the seabed: minerals like manganese, nickel, copper, iron, or titanium that could fulfill global consumption for thousands of years; organic sediments, sands, and placers to be industrially exploited from the continental shelf; and untapped deposits of oil and gas that could be secured from the seafloor's geological layers, liberating coastal nations from foreign energy dependence.[7]

In the mid-sixties, two conflicting imaginaries about the oceans dominated political ambitions, which mirrored the ones previously projected over colonial lands. On the one hand, they were conceived as territorial frontiers facing scrambles to control potential resources. On the other, they were seen as testing grounds for new experiments in international relations.[8] A growing number of historians have addressed how the oceans acquired many of the characteristics typically associated with frontier territories: they came to be regarded as sites to extract inexhaustible resources, areas for industrial capitalism, regions to be grabbed, occupied, and conquered by human communities, and seascapes inspiring future modes of human life.[9] At the same time, the oceans turned into Cold War battlegrounds. Naval powers and coastal nations sought to exert control over ocean space via fisheries in the high seas and as part of their strategy to build up state power in the world order.[10] In the US, national security priorities drove the production of knowledge about the ocean and its seafloor through military patronage over oceanographic research. The practices fostered by state actors around the world were consistent with territorial expansion and control. For example, the SEALAB projects, launched by the US Navy to install underwater habitats, aimed to assert US control of the seabed through its physical occupation. In France, oil firms subsidized the CONSHELF projects, led by the famous Commander Jacques Cousteau, as prototypes of underwater habitation to conduct undersea oil drilling and maintenance activities.[11]

Technological development was at the core of territorial construction and international negotiations. The anticipated uses of the seafloor and its resources fueled innovation in technologies, techniques, and infrastructures; these in turn were the prerequisites to explore, chart, and control the seafloor's potential natural resources. Research technologies such as marine geophysical devices, coring systems, drillships, or diving equipment enabled stakeholders to expand across deeper underwater regions, to penetrate deeper beneath the seabed, to understand with increased detail the seafloor's geological features, and to estimate with higher accuracy its economic potential.[12] Ingenious infrastructures like oil rigs, underwater habitats, or the pillars for amphibious cities were installed on the seabed, reflecting efforts to physically occupy a hostile environment that resisted being tamed. The boundaries of the underwater territory expanded alongside technologies to extract its hidden resources.[13]

The conjoined development of the submerged territory and the technologies to access it were at the core of international legal negotiations that occupied the UN for three decades. For example, at the first UN Convention on the Law of the Sea (UNCLOS) in 1958, the boundaries of the continental shelf—and therefore, the submerged areas under coastal state rule—were established to a depth of 200 meters or "to where the depth of the superjacent waters admits of the exploitation of the natural resources."[14] This left an open door to expand the underwater territory as technological capabilities to access deeper regions evolved. As historian Sam Robinson has indicated, the anticipated uses of the seafloor and future, imagined technological capabilities shaped negotiations at the UN Convention on the Law of the Sea.[15] Frictions arose between those nations that could access the depths and their resources and those that could not but still strove to achieve this ability.[16]

FRANCE AND THE MAKING OF ITS
SUBMERGED TERRITORY

France constitutes a paradigmatic example of how the seafloor became a new territory over which optimistic prospects of economic growth, international power, and national independence were projected. Although rarely considered as such, France is a maritime country whose geography is dominated by the ocean. Its submerged territory today, composed by Exclusive Economic Zones surrounding continental France and its overseas regions, is seventeen times bigger than the sum of those emerged lands. It is the only region in Europe surrounded by three different coasts, all of them of key importance for maritime routes: the North Sea and the English Channel bathe its northwestern coastline, connecting with the Atlantic Ocean on its western side, while the Mediterranean Sea borders France's southeastern shore. Besides the continental space of France (known as the "metropole" or the "hexagon"), the country possesses overseas territories spread across four oceans. French Polynesia, New Caledonia, and the islands of Wallis and Futuna constitute its maritime regions in the South Pacific. Mayotte and the Reunion Islands grant the country access to the Indian Ocean. In the Caribbean Sea, the islands of Guadalupe, Martinique, Saint Martin, and Saint Barthelemy are under France's rule. Next to Newfoundland, in the North Atlantic Ocean, sit the French islands of Saint Pierre and Miquelon, while French Guiana, in South America, opens to the South Atlantic Ocean. These regions make France the second largest maritime territory in the world and together completely determine its political, economic, and military stance in the present, as they did in the past.[17] The oceans have also reflected how France's identity has been imagined, shaped, and reshaped across the last centuries: from the early Mediterranean and the later Atlantic projection, to a global nation, to the crumbling of the colonial empire in the mid-twentieth century that compelled the reevaluation of France's world stance.[18]

Despite its eminently maritime geography, in modern times France has shifted from a transoceanic empire to a continental colonial power.[19] Before World War II, lands under France's control expanded across Africa and the Indian Ocean, Oceania, Southeast Asia, and the Caribbean Sea. The empire collapsed soon afterward. Beginning in 1953, the regions constituting French Indochina in Southeast Asia acquired independence, followed by its territories across the African continent. The process culminated in 1962 with the independence of Algeria, a region central to France's economic and geopolitical strategy and from which France had secured its energy supplies. In short, in the two decades following the Second World War, France faced a daunting series of challenges: to reconstruct a country devastated after the war, to rebuild the social fabric of a demoralized population, and to rethink its strategy of power-building and position in the new world order. Technoscientific development and the taming of natural spaces became tools to rebuild national identity and contributed to the restoration of France's international prestige, such as in the field of nuclear energy through the installation of nuclear plants, or the management of the Rhône River.[20] In parallel, France also turned toward the oceans.

This book tells the story of France and the seafloor, and of how submerged land became an integral part of its national territory via new technoscientific practices and state strategies. It is a story of anticipation in a still unknown environment. The French administration, industrial experts, and its scientists strove to prepare a future economic scenario reliant on the exploitation of submerged mineral and energy resources, long before it was technologically possible and economically feasible to extract them. From the end of the First World War to the late 1970s, the French government's perception of the seafloor (and thus, the importance granted to investing in its exploration) evolved, determined by the futures that were anticipated for the nation as well as the physical spaces upon which its energy security and defense strategy were based. Just before the disintegration of the French colonial empire, the seafloor was a space of mainly military importance for the French Navy, in its pursuit to stand out in submarine warfare and defense. Overseas territories served as bases for controlling the oceans (but also fronts open to external offensives), and government officers understood the importance of deploying fishing fleets as a means to extend power over the high seas. However, after the North African territories sustaining France's energy and mineral supplies obtained their independence, the perception of the seafloor radically changed. For the French oil industry and government officers, the seafloor became a new physical space to secure energy supplies, achieve long-awaited energy independence, and recover France's *grandeur*—the international prestige lost in the wake of the Second World War.

This book's approach combines the history of science and technology with environmental history to delve into the processes and human practices through which the physical, three-dimensional space of the seafloor enters into our worldview, is perceived, and is interpreted.[21] The imagined future uses of the seafloor defined the political perception of this space which, in turn, drove the development of technologies, techniques, and scientific fields that built up the submerged territory—because, whether on land or under the sea, making a territory demands labor, money, effort, and exchanges. Reflecting on parallels between exposed and submerged lands, territorial exploration has historically required coordinated effort, the use of standardized measuring techniques, and the arduous work of traveling across jungles, forests, deserts, airspaces, and oceanic spaces. Data recovered needed to be pooled together and analyzed by different groups of experts, exchanged between nations to define contiguous borders, and eventually represented in maps. Identifying and exploiting natural resources, like underground minerals, required the creation of expert bodies of geologists and geographers, the inauguration of new public institutions, the development of techniques to measure and rule over the underground, and the construction of infrastructures to exploit and transport those resources to processing and consumption areas.

In short, "making" a territory is not just about deploying technoscientific practices to know and tame the environment. It depends on crafting a complex network that connects motivations and priorities at different scales (state, industrial, and scientific), new institutions and public organizations, various funding sources, and actors from different backgrounds. To the difficulties and challenges typical of mainland exploration, the seafloor added the obstacle of doing it *across and through* a mass of seawater—or, as the renowned American marine geophysicist Maurice Ewing bitterly described the ocean, "a murky mist that keeps me from seeing the bottom."[22] The direct experience of this environment is almost inconceivable, and it needs to be mediated by technologies (from the simplest ones, like sounding lines, to the most complex, like drillships).[23]

Tracing the development of research technologies to explore the seafloor implies paying attention to the growth of institutional, scientific, and funding networks of increased complexity. Enthusiastic academic geologists started with the use of dredges onboard modest fishing boats. Later, exclusive geophysical systems and massive scientific drillships relied on international cooperation, industrial support, and large public budgets. Alongside technological and institutional developments emerged new experts in the field of marine geosciences, which encapsulates disciplines and techniques used in the study of phenomena and processes related to the seafloor (from geology, sedimentology, and tectonics to volcanology, geochemistry, and microbiology). In this research, the transformation of technoscientific practices to study the seafloor was profoundly marked, determined, and driven by the French government's enthusiasm to discover and exploit *potential* economic resources from the seafloor. As future interests in the seafloor took shape on the state's agenda, diverse groups of actors—geologists, oceanographers, government officials, and oil industry experts—coalesced around a shared goal: to know the seafloor. Different sources of public funding converged around these groups of people at the same time that state officers designed mechanisms to drive scientific research toward the exploitation of marine resources. This process culminated in 1967 with the creation of the National Center for the Exploitation of the Ocean (the Centre National pour l'Exploitation des Océans, or CNEXO), an oceanographic center that played a major role in defining France's foreign oceanographic policy. In other words, France transformed the exploration of the oceans and their seafloor into a means to reaffirm its position in the postwar world order.

A TALE OF TWO COASTLINES: DECENTERING NARRATIVES OF OCEAN(FLOOR) SCIENCES

By focusing on France and its evolving relation to the seafloor, this book contributes to the recent history of the oceans by decentralizing a narrative of exploration and exploitation that has largely concentrated on the US and on military patronage. For the last two decades, historians of science, technology, and the environment have produced an extensive literature showing how, following the end of World War II, the US came to dominate ocean exploration through the marine sciences. American research institutions and oceanographers led the development of research technologies, methods, and cooperative projects, while federal administrations launched national programs of oceanic development and utilized ocean research to craft diplomatic relations.[24] In these historical accounts, military concerns, priorities, and funding stand out as the drivers for seafloor exploration and the development of marine geosciences.[25] In particular, historians of science Naomi Oreskes and Jacob Hamblin have accurately characterized the nature of American marine (geo)sciences in the aftermath of World War II, showing how the regime of military secrecy shaped knowledge production about the seafloor, and how international and military partnerships became two sides of the same coin in cooperative scientific surveys and projects. Together, their works encapsulate a robust body of academic work from which we, historians of ocean sciences in the Cold War, have drawn our understanding about the development of marine geology and geophysics. But should we infer that marine geosciences developed under the same motivations, pressures, and through parallel currents in other regions as in the US? And if not, what might these differences reveal about our understanding of the oceans' recent history?

The question, as simple as it sounds, arose as I first approached the history of marine geosciences in France. The starting points of this research were stories written by French geologists who combined their firsthand experiences in the nascent field with the scientific results, the tools used, and the social relationships established.[26] As framework and reference point,

I relied on the historiography of Cold War ocean sciences in the US. Yet I soon realized that the origin of seafloor sciences and technologies, in France, had moved along a different pathway. Formal interviews with French geologists quickly revealed that the oil industry emerged as a major partner to academic researchers. A fortuitous encounter with the (massive) archival collection of CNEXO at the National Archives of France opened up an entirely new perspective on the recent past of seafloor exploration. Reports, letters, and minutes revealed imagined offshore industries, unrestrained economic ambitions, high-level diplomatic relations with the oceans at their core, and a new scientific organization to fuel ocean exploration. A later visit to the Defense Historical Service (Service Historique de la Défense) confirmed the hypothesis: while military concerns and priorities certainly underlined the exploration of the seabed, an unprecedented governmental ambition to stimulate an ocean economy and restore France's prestige through ocean sciences propelled the exploration of the seafloor in its three dimensions.[27] In other words, to develop our understanding of how the seafloor has been approached, tamed, and shaped, we need to integrate the narratives of nations other than the United States or Britain—nations that held distinctive geopolitical priorities, economic needs, and territorial settings—as well as to include patronage relationships beyond the science-military alliance.[28]

The stories that emerge from these multiple perspectives tell a more complex and interwoven history of the seafloor. To point to a widely known example: plate tectonics theory, which completely altered the way in which we conceive Earth, resulted from research mostly conducted in American and British research centers, relying on data acquired through military funding or surveys. We now possess a solid literature, authored by scientists, historians, and social scientists, on the processes that led to the theory's materialization.[29] But there was a less-explored sequel. Plate tectonics theory worked like a new pair of eyeglasses for Earth scientists, through which emerged *and* submerged mountain chains, plains, and valleys were seen as never before. Existing data were insufficient, and older publications became outdated. Around the world, geologists eagerly sought to go to the field—*and* to the sea. The ocean crust transformed into the epicenter for understanding Earth's dynamics. Against this backdrop we should ask, How did different regions and countries contribute to our understanding of the world's ocean floor and its dynamics? Which motivations and patronage relations fueled their ships and fed their laboratories? And how did those differences shape the technologies used and the knowledge produced?

In this book, by emphasizing, in a distinctive country like France, economic motivations as drivers and industrial organizations as scientific patrons, I seek to engage with an emerging scholarly conversation that diversifies the actors that have played a role in producing our current knowledge about the oceans.[30] Note that decentering narratives does not mean disconnecting them. To the contrary, in France, ocean exploration thrived in intimate relation with US researchers, institutions, industries, and government policies. For the oceanic history of France, the US-centered historiography is the main reference point, because the geologists, industrial experts, and public officers that populate these pages also looked to the other side of the Atlantic. As the reader will notice, each chapter features a US-based researcher, underwater technology, institution, or geopolitical event. These constituted crucial triggers for the development of France's oceanic policy, providing a role model (and on occasion, an adversary) to advance oceanographic capabilities.[31] In this sense, this books aims at establishing a conversation with the histories of other oceangoing countries, because that is the only way we will grasp with fine detail and precision how our understanding of the seafloor was shaped and how, in turn, this shape is reflected in today's open issues of oceanic governance, economic expectations, and cultural representations.

INDUSTRY, MILITARY, AND ACADEMIA: A NOTE ON CHOICES AND APPROACH

Four remarks are needed on the choices and approaches to extractive enterprises in France, underwater military surveillance, and the research activities that this book does (and does not) follow. For its development and characteristics, the study of France's oil industry must be addressed as an integral part of the state system, rather than as a composite of private firms. The origins of the French oil industry can help explain this. For France, acquiring energy independence became a national priority in the aftermath of World War I. The Allied powers had "floated to victory upon a wave of oil," as British member of the War Cabinet Lord George Curzon famously asserted to account for the relevant role American and British energy supplies had played in winning World War I.[32] Meanwhile, France had suffered serious challenges to securing its energy supplies, because it lacked producing grounds and infrastructures of its own. Conscious of the need to establish a national petroleum policy, the French government began to build its own oil industry during the twenties.[33] Its intervention in the energy sector was remarkable during the following decades. The oil firm *Compagnie Française des Pétroles* (CFP), of which the French government owned third, was created in 1924 to manage the French share of the Turkish Petroleum Company in the Middle East.[34] The aim of exploring and producing hydrocarbon supplies in the national territory—underground continental France, across North Africa, and in controlled regions of the Middle East—stimulated the creation of other companies and oil-related institutions, which were also completely or partially subsidized by the French government. The Régie Autonome des Pétroles was a national firm created in 1939 to exploit gas deposits in southern France; two years later, the firm Société Nationale des Pétroles d'Aquitaine (SNPA) was inaugurated, with the state as one of its two shareholders. In 1945, the Ministry of Industry created the Bureau de Recherche de Pétrole to coordinate hydrocarbon exploration among different French institutions, which in later years acquired a relevant role in exploring the resource potential of Saharan colonial territories.[35] These are only a few examples to illustrate that, instead of a collection of independent private firms, the French oil industry was composed of institutions completely or partially subsidized by the state. Because of its dependence on government, its connection to state goals, and its funding system, the funding structure of France's oil industry resembled military patronage more than funding relations among completely private corporations. In short, in France the state was the main actor both in the military domain and in the oil industry. For this reason, this book does not treat the oil industry as a private enterprise but rather as an integral part of the state system.

Also related to commercial activities, the reader may have already noticed in this book the repeated use of the word *exploitation* as an equivalent to *production, extraction,* or *use* of natural resources. While in modern English the term often carries a negative connotation—implying abuse, overuse, or inequitable advantage—I have chosen it deliberately, as it was the predominant term in discussions of ocean resources during the period covered in this book. It appears consistently in UNCLOS treaties, including those still in effect, and it even featured in the name of France's oceanographic center, the National Center for the *Exploitation* of the Oceans.[36] Rooted in the Latin *explotare* (to make use of, to produce a result) and borrowed from French into English, its negative connotations in English probably emerged in the US antislavery writings of the 1830s through the 1850s.[37] In contrast, in French, it generally retained a neutral meaning of resource production and use. This may explain its persistence in international discussions on ocean resources where both languages were often used (for example, in 1902 the International Council for the Exploration of the Sea was established to ground the "rational exploitation" of fisheries on scientific results).[38] By the 1970s, however, rising ecological awareness, postcolonial debates, and sustainability concerns progressively added a critical dimension to the term for marine and emerged environments. I use *exploitation* here to reflect the language of the time, while remaining mindful of its con-temporary dual connotations of use and overuse, production and harm.

Although this research focuses on the exploration of natural resources and the *economic* motivations projected onto the seafloor, military inter-ests and concerns intersected. During this period, France first developed its submarine capabilities for nuclear deterrence and built up its surveil-lance strength, propelled by its position vis-à-vis the North Atlantic Treaty Organization (NATO). Despite being a founding member of the alliance, France's relation to it was rather tense—especially during Charles de Gaulle's administration, from 1959 to 1969. In 1964, President de Gaulle withdrew the French Mediterranean fleet from NATO, and two years later withdrew from NATO's integrated military command. France's military priorities revolved around its autonomous control of the oceans. The coun-try needed to defend more than 3,000 kilometers of coastline in the hexa-gon alone, together with its new and vast underwater territory, which was being delimited at exactly the same period by the United Nations' Conven-tion on the Law of the Sea. Such parallel development spotlights questions about the relationship and codependence between the seafloor's economic exploration and the ocean's military surveillance, especially considering that both—the military and the industry—were branches of the public ad-ministration.[39] It is likely that France's strategy to control the oceans mili-tarily relied to some extent on its economic exploration and vice versa. For example, some military surveys (like NATO's cooperative programs, or the topographical data acquired by the French Navy's Hydrographic Service) ended up providing useful information to identify natural resources. And some archival hints point to a convergence of interests in France's explo-ration of polymetallic nodules in the Pacific and military surveillance of the region. Yet the true depth of the connections between the military and offshore industries in seafloor exploration, in France as in other countries, awaits exploration.

Finally, a remark on the featured research teams and institutions in France. This book follows individuals who engaged in seafloor exploration to fulfill national goals, like Jacques Bourcart. It also traces the develop-ment of particular institutions, organizations, and devices that were central in the state's program of resource exploration. Beginning with chapter 4, CNEXO becomes the main narrative actor as a newly designed state tool to steer ocean research toward geopolitically relevant goals. Just like human actors, institutions have a life, a birth, and a point of demise, marked by changes in their structure, governance, or direction. Scientific institutions have power to set processes into motion, drive relationships, and represent interests that are beyond the science there developed. The difficulty is to infuse blood and life into an institution—or to give visibility to the sum of individual actions that build up the institution's actions. CNEXO's featuring role should be understood as the combined activities of a collective of actors who, through their discussions, decisions, and efforts, maneuvered the institution's movements: government officials, industry representatives, ocean stakeholders, and researchers from different fields, sitting together at the Administrative Council, discussing in closed-door offices, or planning marine projects at laboratories. CNEXO's importance lies in that it was conceived as an underwater equivalent to organizations for space exploration and nuclear research—the technoscientific fields driving international relations in the Cold War. It became a national hub for pioneering and richly supported research activities, the managing center for France's oceanographic fleet, and the official representative of France's ocean activities internationally. But beyond CNEXO's research team, French universities around the country hosted departments and research groups devoted to marine geology and geophysics whose research was not directly supported by CNEXO—although they depended on CNEXO's agreement in order to plan their cruises or use national oceanographic resources. In other words, this book is not a complete history of marine geology and geophysics in France, nor does it intend to map exhaustively all French laboratories and research teams that investigated the seafloor during the sixties and seventies. Neither does it follow the development of all seafloor research technologies (the evolution of submersibles or instruments for conducting magnetic studies, for instance, are not central in this narrative).[40] Rather, it focuses on those instruments commonly used by both academic researchers and offshore industries, whose stories shed new light on the interplay between science and industry in the production of knowledge about the seafloor—eco-sonars for charting the seabed's topography, marine geophysical devices for reflection seismics (applied to "read" the seafloor's layered composition a dozen meters beneath the seabed), and ocean drilling (to recover samples from the deep seafloor).

Exploring the seafloor's economic potential was only one of CNEXO's priority areas—although it is the one at the center of this book. The institution's research agenda was divided into five interdisciplinary topics, each one addressing an embracing problem: exploitation of mineral and fossil resources (which included seafloor exploration), exploitation of living resources, coastline management, control of pollution (oil spills, urban discharges, etc.), and the ocean's interaction with meteorology and climate conditions. Even a brief glimpse of the list can surprise the reader: How could two apparently contradictory priorities, exploitation and conservation, coinhabit the agenda?

During the 1960s, numerous authors, explorers, and oceanographers inspired the exploration of the oceans as a subject of popular fascination through publications, TV shows, and films.[41] This popular perception combined understanding oceans as infinite sources of natural resources to sustain human life with a slowly growing environmental concern for the negative impacts of human activities. In the US, environmental concerns bloomed after the success of Rachel Carson's book *Silent Spring* (1962), which constituted a turning point in environmental history as it opened a dialogue about the relations between humans and nature, public health, and conservation movements. Greenpeace was funded in 1967 to disrupt nuclear tests in the South Pacific.[42] In France, the popular awareness of the ocean's fragility began to emerge due to the environmental disaster of the *Torrey Canyon* oil spill on March 18, 1967, just weeks before CNEXO was inaugurated. The tanker crashed thirty-two kilometers off the English coastline of Cornwall, spilling more than a hundred million liters of oil that covered beaches across the coastlines of France, northern Spain, and the UK. The incident was the germ of a growing environmental movement in France which, first timidly and then, by the eighties, more vividly, strongly opposed hydrocarbon-related activities at sea.

The apparent contradiction of embracing the exploitation of the oceans as an endless cornucopia, even as environmental consciousness rose across the globe, can be explained by the terrestrial bias of environmental movements. According to historian Helen Rozwadowski, by the mid-sixties, conservation was focused on preserving land environments like forests, mountains, lakes, or plains; and action was taken to prevent pollution and protect public health. For the oceans, however, the solution to recurring problems—like overfishing, or waste dumping—were to be solved through improved scientific knowledge: marine biologists could help identify new fishing grounds and new economically valuable species, while marine chemists still believed that the solution to pollution was dilution. The vast ocean, beyond the coastal zones, was seen as practically immune to human influence, and the seafloor, as a yet inaccessible El Dorado.[43]

FROM THE SHORE TO THE DEPTHS: CHAPTER OUTLINE

The chapters follow the expansion of seafloor exploration chronologically, in extent and in depth, through expert networks and France's ambitions to discover the oceans' riches. Geographically, the book opens with the study of the Mediterranean basin, starting in its coastal and shallow regions, progressively expanding the analysis to more distant and deeper areas, involving technological innovation, larger sums of funding, and the new expertise and collaborations that enabled French experts to reach deeper into the Atlantic and Pacific Oceans. The final chapter returns to the Mediterranean, illustrating how the conduct of research on the seafloor had changed dramatically in less than three decades.

"Deep Blue Canyons: Geology from Land to Sea," the first chapter, spans World War I to the late 1950s, when the first coastal laboratory for marine geology was established in the French Mediterranean. It explores how the seafloor's perception as a submerged, unexplored, and rich territory emerged during the world wars, driven by military advancements in underwater technologies and culminating in the 1945 Truman Proclamation on the continental shelf. Divergent geopolitical strategies on energy security in France and in the US shaped their views on the seabed and thus on their different support for marine geology. While the former secured hydrocarbon supplies in colonial territories, the latter turned to the oceans. The narrative focuses on Jacques Bourcart, a French geologist who directed the shift from mainland studies to the continental shelf. A frequent collaborator with the French military, Bourcart's career illustrates how France's military priorities shaped his geological pursuits on land and underwater. His interest in submarine canyons converged with the research of Francis P. Shepard, Bourcart's American counterpart. The comparison of their careers reveals divergent pathways in the development of seafloor exploration. While in the US the oil industry's support enabled complex marine geological programs, in France, the navy's patronage fell short in fostering the discipline's growth below the surface of the continental shelf.

Chapter 2, "Lagging Behind? Geopolitics and Geophysics Underwater," tracks the oceans' progressive integration into France's political worldview, conceived as scenarios for deploying geopolitical strategies and rebuilding national prestige. In 1957–58, three key events spurred this shift: the International Geophysical Year (IGY) strengthened the link between national prestige and international oceanographic collaboration, the first United Nations Convention on the Law of the Sea emphasized coastal nations' role in ocean governance, and Charles de Gaulle's inauguration framed technoscientific development as means to restore France's *grandeur* and mend national identity. These events exposed France's limitations in ocean exploration, prompting the inclusion of marine sciences in the government's priority agenda. This novel political perspective opened up a new dimension of the seabed: the vertical one. The chapter follows the introduction of marine geophysical technologies to the toolbox of French geologists to survey the Mediterranean, backed by an unprecedented public budget. The parallel development of marine geophysical devices in the US shows that, while French geologists were inspired by their American counterparts in their research choices, significant disparities in research efforts persisted on both sides of the Atlantic. The feeling of "lagging behind"—used in political and scientific contexts—intensified in 1965, when American researchers made unprecedented discoveries in the Mediterranean seabed.

"France's New Economic Frontier," the third chapter, shifts the focus of attention from academic geologists to the oil industry, revealing how the independence of North African colonial territories transformed the seafloor into "virgin regions" for industrial exploration and exploitation.[44] Starting with Algeria's independence in 1962, the narrative turns to follow André Giraud, the civil servant at the Ministry of Industry who orchestrated the reorientation of France's oil industry toward the oceans. Technological industrial development, exemplified through marine geophysical techniques, was soon deemed insufficient to speed up the seabed's economic assessment. Close cooperation with academic researchers became imperative. The chapter culminates with the spill of industrial priorities into France's oceanographic agenda, describing one of the most relevant turning points in the recent history of France's ocean exploration: the creation in 1967 of the National Center for the Exploitation of the Oceans (CNEXO), a government-led structure to steer technoscientific development toward economic revenues. The chapter closes by introducing Yves la Prairie, former naval officer and CNEXO's first general director, whose leadership shaped the future of seafloor exploration.

Chapter 4, "Three-Dimensional Territories: Science and Industry in the North Atlantic," delves into the convergence of scientific and industrial priorities in joint oceanographic surveys led by CNEXO. The blending of interests, technologies, and experts played a central role in shaping the conception of a three-dimensional, underwater territory. This underscores that the industry's contribution was fundamental and inseparable from our current understanding of the seafloor—while scientific interests informed economic evaluations. The marine geophysical projects organized during CNEXO's initial years (1968 to 1972) reflected the political ambitions associated with marine geosciences. The cruises Noratlante I, Nestlante I, and Nestlante II contributed to the double purpose of enhancing France's international profile (by entering into plate tectonics research and stressing its presence in the high seas) and supplying crucial data to the oil industry consortium. The only large oceanographic vessel in France, the *Jean Charcot*, became the common ground where industrial interests in surveying continental margins merged with scientific motivations to study the dynamics of the ocean crust.[45] In parallel, the national cartographic program on France's continental shelf simultaneously revealed the geological history of

the Atlantic basin and outlined its resource potential. CNEXO not only succeeded in creating pathways for industrial-academic collaboration but also
demonstrated its key role in constructing the submerged territory where
the French extractive industry could operate in a near future.

While chapter 4 describes the national industrial-academic network
that CNEXO built to explore continental shelf and margins in France and
nearby regions, chapter 5 moves from the national to the international setting to explore how, in parallel, the institution expanded its activities to the
deep Atlantic and Pacific Oceans. "Alliances and Hidden Minerals in the
Abyss" looks at how commercial interests and international mottos shaped
the kind of relations that France established to cooperate in exploring the
deep ocean floor. In the late sixties, mining deep-sea minerals beyond national seawaters was still a prospect, but within the United Nations, several
countries expressed concerns about potential inequalities that could arise
from differing national technical capacities. These fears mixed with other
ideas that were gaining momentum within ocean diplomacy: collaborating for the benefit of humankind and a race-like competition to identify
economic resources. For France, taming and controlling the deep ocean
floor through marine geosciences required new technologies, like submersibles and mining instruments, but also the establishment of international
relations and agreements to avoid potential frictions. This chapter analyzes
CNEXO's science diplomacy role by looking at its activities to explore two
scenarios, the Atlantic and the Pacific ocean floors. For each, it describes
the creation of a CNEXO research base and deep-sea research program.
The Franco-American collaboration illustrates CNEXO's pursuit to explore
the deep Atlantic, in a relation permeated by a constant dynamic of competition and collaboration. The Oceanological Center of Brittany, conceived
as an international base for transoceanic cruises, was openly inspired by
American research centers, symbolizing a token of mutual trust between
the two countries. The Franco-American Mid-Ocean Undersea Study (FA-
MOUS) served as a display of their joint technological prowess under the
mid-Atlantic ridge, gathering key data that proved seafloor spreading while
also hinting at the existence of mineral deposits. For the exploration of
polymetallic nodules in the Pacific Ocean, the chapter details how CNEXO
established relations to gather information from foreign research centers
and industries while also building up a logistical base in Polynesia for international research and mining tests.

Chapter 6 returns to the Mediterranean two decades after Bourcart first
explored its continental shelf. "Stories Beneath Deep Salt: Drilling Across
the Mediterranean" focuses on the discovery of a thick layer of salt buried
beneath the deep Mediterranean seabed, using groundbreaking technology:

the scientific drilling vessel *Glomar Challenger*. These salt rocks, composed of evaporitic minerals, attracted both the international geological community and French industry experts. For scientists, they held clues to the Mediterranean's distant past. For industry, they promised hidden hydrocarbon reserves near continental France. The chapter explores how international science advanced deep oil assessment, while commercial data drove scientific progress. In this sense, it conveys a defining message of this book: advances on seafloor research cannot be understood without acknowledging the contributions of commercial expertise, data, and technology. What sets this story apart from earlier ones is France's role in a scientific initiative where it was neither a leader nor a funder, but one invited participant. The *Glomar Challenger* operated under the Deep Sea Drilling Project (DSDP), an American project launched in 1968, funded by the National Science Foundation (NSF), and led by American experts. Yet France, through CNEXO and its combined scientific and industrial expertise, found ways to maximize its involvement. The French oil industry grew interested in the DSDP for two reasons: visions of future deep-sea production and fears of American economic discoveries in the western Mediterranean. French oil geoscientists integrated into the international community, contributing to expedition planning and postcruise analyses. Their collaboration helped shape a field of research on the Mediterranean's past whose scientific significance remains relevant today. As well as this American program, its technology and findings are an integral part of the history of France's underwater territory; so were commercial interests in one of the Mediterranean's most important geological discoveries.

We don't usually think of the seafloor as a territory, but it is a central arena for today's global governance, economic forecasts, diplomatic relations, and social functioning. We don't think of France as a maritime country, yet the management of its underwater territory permeates every political action, strategy, decision, and relation. This book is a call to historicize the seafloor, to approach the underwater from a different perspective, and to integrate land below sea within narratives of national governance, international relations, and sciences. Only by delineating the contours of the seafloor's past will we see the influence of its shape in today's world.

CHAPTER 1

Deep Blue Canyons:
Geology from Land to Sea

The scientific conquest of the world is something from the past. The
very word "exploration" prompts smiles. . . . Yet, under the mass of
water that covers 72% of the globe, there is still a vast expanse of lands
to discover!

J. BOURCART, *LE FOND DES OCÉANS*

Imagine we are standing at the edge between the sand and the sea, watching
the subtle coming and going of the waves breaking off the beach. From the
wet sand beneath our feet to the horizon, the ground first visibly extends
underneath the clear layer of shallow seawater, then disappears below the
dark blue surface that covers the entire landscape in front of our eyes. Unlike a terrestrial landscape, the contours and shapes, mounts and valleys of
the seafloor are invisible, and almost unreachable.[1]

The story of the seafloor's exploration starts from land, powered by geologists who imagined revealing the geography hidden beneath the mass
of water. In France, Jacques Bourcart was among the first who shifted his
perspective from emerged to submerged lands. Lecturer in geology and
geography at the University of Paris, Bourcart was a professor who was
both adored by and unpopular with his students. Always with a cigarette in
his mouth, his lectures often transformed into never-ending stories of his
firsthand experiences on the front during World War I. But he could also
talk for hours about the mysteries of the only frontier humans had never
reached—the deep seafloor—with an emotion that attracted disciples. His
academic colleagues considered him an eccentric researcher with bizarre
scientific ideas; yet the first generation of French marine geologists were
trained under his charismatic mentorship. Bourcart wove together mem-
ories of the landscapes that he had explored in his youth with vivid images
of the geography he imagined underwater. As his praise for the idea of exploration hints, Bourcart devoted part of his career to exploring, mapping,

and understanding foreign territories for the French army. His experience as a geologist on land was echoed in the new era of adventures that awaited beneath the oceans. His younger self had pictured the ocean floor as a vast desert akin to the Saharan plains he had traversed with a cavalry brigade during World War I; later he suggested underwater geography was much more complex, perhaps similar to the Atlas Mountains in North Africa. Just as physical geography is crucial to shaping human life, communities, and historical processes—lessons he learned both in his studies and then first-hand while traveling through wartime Albania—Bourcart was certain that, beneath the seawater, the seabed's geography similarly determined Earth's history.[2] In particular, he dreamed about unveiling the natural history of the Mediterranean basin, intrigued by a geological event that, according to his investigations, could have transformed entire submerged regions into exposed mainland areas. Geologists termed this a marine regression, an event during which the sea level lowers dramatically due to movements of Earth's crust.

To understand how the seafloor has been explored, charted, and known, this book begins with the turn of the twentieth century when some geologists, like Jacques Bourcart, began to sample rocks, sands, and gravels from mainland areas that were once submerged under the sea. Bourcart's academic career exemplifies not only the step from mainland exploration to the offshore but also offers a lens through which we can understand how key transformations in the international context (colonialism, two world wars, new geopolitical strategies) shaped the view of the oceans. Small and regional as it might seem, Bourcart's step into the oceans represents an event comparable to the first step of men on the moon: the opening of a pathway for exploring a new territory via technological innovation, supported by military institutions, extractive industries, and national governments eager to anticipate its future uses. The seafloor's economic imaginary began to emerge at the same time its geological understanding came into focus.

Yet we should not simply take the regional as universal. The development of marine geosciences in France followed a different pathway than that experienced elsewhere. This chapter moves from France to the US, alternating between the development of seafloor exploration in the two countries, by following the biographies of Bourcart and Francis P. Shepard, his American counterpart and intellectual reference. The careers of both men illustrate how different political priorities, strategies, and responses to global events led to different understandings of the seabed's geopolitical importance and, thus, to an asynchronous development of marine geosciences. In short: In the US, the aftermath of World War II brought in a new vision of the near-shore seafloor as a promising hydrocarbon supplier, to survey and secure by developing marine geosciences. This political outlook developed at a later stage in France, since the country planned to secure energy supplies on North African colonial territories. Bourcart's attention to American researchers symbolizes the beginning of a constant comparison of research capabilities that, in the following decades, would find a parallel in France's governmental bodies, industries, and scientists.

THE SEAFLOOR FROM THE MAINLAND

Jacques Bourcart, born in the small Alsatian village of Guebwiller in 1891, grew fascinated by the sea during his childhood, when he spent summers with his family on the French Atlantic shore. This passion drove him to pursue undergraduate studies in natural sciences at the University of Paris and, in 1909, to engage in marine biological research at the Parisian Institut Océanographique.[3] However, his training in ocean sciences proved short-lived. In 1912 Bourcart was called up to perform his military service in Morocco, where he remained in a cavalry brigade during World War I. Military activities led him to redirect his academic career to geography and geology on the mainland, although he never forgot his curiosity about the oceans. Rather, he ended up devoting his entire career to studying past transformations of the Mediterranean Sea by analyzing its seafloor geology, starting from the dry land.

Bourcart began his journey of seafloor exploration far from the sea, on the Albanian front, where he was called to serve in 1917. One year before, French regiments had occupied Korçe, declaring it an autonomous region under France's administrative control due to its strategic value—the area connected the Italian front in Albania with the French in Macedonia, isolating the Greek region under Austrian-German control.[4] In the French Army, naturalists were considered valuable experts for exploring and charting enemy territories, so the Geographical Service of the French Army entrusted Bourcart to gather strategic information to effectively deploy French troops, including on Albania's geographical features, the condition of roads and paths, the amount of local resources available to supply French troops, and the local population's character.[5] During these military reconnaissance missions, Bourcart took the opportunity to make geological studies of the region, sampling rocks, drawing observations, and investigating the fossil microfauna he found. The seemingly peaceful activities of a naturalist, quietly observing and drawing the landscape, alternated with the frenzied situations characteristic of the war's front. For example, although he lacked medical training, Bourcart worked as an auxiliary doctor when wounded soldiers returned from the battle and, at some point, he was severely poisoned with mustard gas.[6]

Albanian outcrops displayed sediments corresponding to a deep marine environment, abruptly followed by sediments that tend to accumulate in dry, mainland regions. Bourcart asserted that this sequence could only be explained by a vaster Mediterranean Sea once covering the region of Albania, merging the Aegean with the Adriatic Sea. He suggested that a subsequent retreat of seawater led to the emergence of the territory we now know as Albania.[7] From his in-depth understanding of Albania's geology and its changing coastal profile, Bourcart began to conceive of the Mediterranean basin as a structural unit, where geological events would have occurred simultaneously across the whole basin, shaping all coastal regions. But in order to prove it, he needed to travel to the other corner of the basin: North Africa. For that ambition, major changes in the global geopolitical context in the wake of World War I proved particularly fertile.

At the war's end, it became evident that the way of conducting war had changed: the introduction of fast, highly mobile vehicles, like airplanes, tanks, large battleships, and submarines not only modified military strategy but also transformed their fuel—oil—into the most valuable resource worldwide. Between 1910 and 1918, global production of oil doubled, increasing from 40 million to 80 million tons per year.[8] During the war, France had suffered serious supply difficulties, since the routes through which Russian and Romanian oil reached the country felt under Ottoman control. At that moment, France had turned to American and British producers as its future suppliers. When the war ended, and after witnessing the pivotal role an abundant oil supply had played in the Allied victory, the French government established as its national priority the development of its own petroleum industry. Aiming to secure oil supplies for the country without relying on Anglo-American companies, the French government created the national *Compagnie Française des Pétroles* (CFP) in 1924.[9]

French territories appeared promising for oil exploration, as France's colonial empire expanded across the globe—and the oceans. Its extent was only surpassed by the British colonial empire. On the African continent, fifteen regions were under French rule from the mid-nineteenth century on, including Algeria and Tunisia.[10] In Asia, the French Indochinese Union and French establishments in India rested under its rule until 1954, as well as some dependencies in China. In the Middle East, regions in Syria and Lebanon became France's protectorates starting in 1920. And France controlled numerous islands and archipelagoes in the Indian Ocean, the South Atlantic, and the Pacific Ocean—some of those regions still remain part of France's overseas territories today.[11] In this framework, trying to secure energy supplies from these territories appeared a logical strategy, one that would provide stability to continental France (fig. 1).

Bourcart could resume his investigations in colonial Africa through newly government-founded institutions to coordinate oil exploration and surveying throughout Algeria, Morocco, and sub-Saharan Africa. His results, in turn, contributed to building a strong national oil industry.[12] The Service of the Geological Chart of Morocco, a national organization to gather militarily strategic information and analyze the economic potential of the Moroccan geography, financed Bourcart's missions across North Africa.[13] The mining potential of those territories was completely

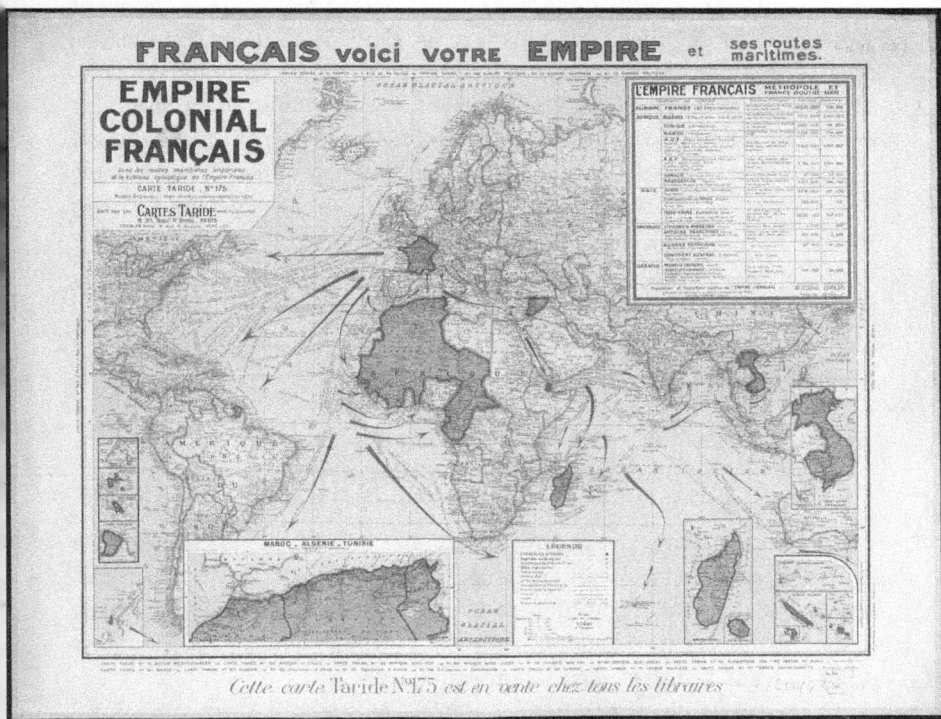

Figure 1. The French colonial empire in 1938. The heading reads, "French people, here you have your empire and its maritime routes." The map is a Eurocentric representation of the globe's ocean and continents, highlighting only France's territorial possessions. It transmits the sense of a shrunken world that orbits around France.
"Empire colonial française avec les routes maritimes impériales et le tableau synoptique de l'Empire français, 1:34.000.000 à l'Equateur," Taride (Paris), 1938. Bibliothèque Nationale de France, Département Cartes et Plans, GE CC-391 (175). Virtual access through the library's Gallica digital repository, public domain. Reprinted with permission of Bibliothèque Nationale de France.

unknown, hence military and industrial goals converged toward a single priority: establishing a detailed geological cartography of the upper sediments, which was Bourcart's main task.

Moroccan geology and its fossil evidence revealed a similar event as the one that had shaped Albania's geography: an underwater landscape that became suddenly arid, characterized by fossil beaches and sand dunes.[14] Such a rapid change of paleo-environments could only be explained, in Bourcart's mind, if the continental mass had moved vertically, pushing the seawater to retreat to the deeper regions of the basin. As fascinating as it is to imagine the seabed suddenly emerging from beneath the water, Bourcart's ideas did not find much acceptance within the French geological community. Minor oscillations of seawater levels were accepted as frequent events that, throughout Earth's history, had shaped coastlines, but the idea of an abrupt marine regression of hundreds of meters was unimaginable. This lack of acceptance among his scientific community made Bourcart aware that he could not continue supporting his ideas with mainland evidence alone; he needed to get his feet wet, to go into the sea and sample the seabed. His determination was inspired by American geologists, who were already moving their geological ventures below the ocean's surface by applying new military-based technologies. Their underwater discoveries made Bourcart realize that, if the same evidence was also found in the Mediterranean continental shelf, he would have some proof for a massive seawater retreat.[15]

TOOLS TO FATHOM THE DEPTHS: A CONCISE HISTORY OF ECHO-SOUNDERS

To understand the striking new ability that military technologies gave American geologists in seafloor exploration after World War I, it is necessary to first understand how the seabed was fathomed previously, going back to the previous century, and the processes through which echo-sounders originated and transferred to oceanographic research.

The ocean depths were discovered in the nineteenth century, marking the start of the human relationship with all the ocean and its seabed. Before that era, only navigators and some curious naturalists had attempted to carefully measure the depth of shallow waters along the coastline, using sounding lines of about 100 fathoms (less than 200 meters). To identify the nature of the sediment for fishing or anchoring purposes, navigators covered the lead situated at the rope's extreme end with sticky materials like tallow. The open ocean, however, was deemed unreachable or even bottomless. It was not until the advent of the telegraph that governments began measuring the depth of the seabed systematically. In 1858, the first transatlantic telegraph cable achieved communication between the American continent and the UK. Sounding the seabed became, first, the responsibility of government hydrographers and, eventually, the routine work of cable company employees. Bathymetrical charts of the North Atlantic were initially published for navigational purposes, but they also came to delight a population that was becoming acquainted with, and fascinated by, the seas and the oceans.[16]

After the turn of the century, a new system to fathom the seabed's depths and topography was envisioned: echo-sounders. These instruments relied on the same physical principle that enables bats, toothed whales, and dolphins to communicate—a phenomenon not yet understood at the time. By emitting and receiving a sound wave, these animals obtain information about the distance and size of surrounding objects, through which they can navigate underwater or in the dark. Echo-sounders were first invented after the tragic sinking of the British passenger liner RMS *Titanic* on April 15, 1912. In order to prevent similar accidents, German physicist Alexander Behm focused on creating a navigating system that could detect icebergs, but his prototypes failed repeatedly when tested due to weather conditions—the effects of salinity and temperature changes on sound velocity were poorly understood. But in February 1916, when testing a new system in the shallow waters of the Baltic Kiel Fjord, Behm discovered another application for his devices: to sound the seabed. Using explosives as a sound source, Behm generated sound waves strong enough to travel down to the seabed, there ten meters deep, and return to the ship's receiver. The interpretation of the waves' velocity provided numerical data on the seabed's depth and, thus, on its topography when sounding was extended across larger spans.

Behm was not the only expert testing underwater systems for navigational purposes. In parallel to his investigations in the Baltic, Canadian-born inventor Reginald Fessenden was undertaking similar experiments in the US. Working as a consulting engineer with the Submarine Signal Company, Fessenden designed the first prototypes to enable communication between vessels and mainland stations in 1912. By ringing underwater bells, sailors could send about twenty Morse-code words per minute; these were received by hydrophones up to eighty kilometers away. He triumphed over Behm in developing a functioning system to detect icebergs up to twenty kilometers away from a vessel. Like his European competitor, Fessenden also designed an echo-sounding system that could replace lead lines and other rudimentary techniques to fathom the ocean's depths and chart the seabed.[17]

The outbreak of World War I stimulated the interest of combatant nations in underwater acoustic systems because of their ability to detect enemy submarines and to enhance communication between allied vessels while at sea. The US equipped its warships with the so-called *Sub Sig* system, although with little success, while the British navy designed prototypes that worked at higher frequencies.[18] The French Hydrographic Service conducted similar experiments, during which hydrographer Pierre Marti developed a prototype using firearms: the explosions of the cartridges generated the sound waves used for depth sounding. Although peculiar, the system proved to be successful, enabling the Hydrographic Service to sound the Bay of Biscay from shallow waters, less than 200 meters deep, to the open ocean, up to 4,000 meters deep.[19]

When the war ended, private companies and naval organizations improved echo-sounders, and their use spread as well among oceanographic laboratories. During the twenties and thirties, national governments, private institutions, and philanthropic individuals supported numerous oceanographic campaigns around the world, refitting military vessels with echo-sounders to explore the depths and unveil the topographical features of the seabed while conducting simultaneous investigations into marine biology and physical oceanography. Just to name some examples, the German *Grosse Atlantische Expedition* (widely known as the Meteor Expedition for the name of its vessel, 1925–27), the US *Carnegie*'s Cruise VII across the North Atlantic and Pacific Oceans, the British Antarctic Expedition led by explorer Richard E. Byrd (1928–30), or the Danish circumnavigations of *Dana I* and *II* (1928–29) were among the interwar oceanographic expeditions that provided unprecedented knowledge about the oceans. Among the most relevant results of a thorough sounding of the seabed, their discoveries together demonstrated that the continental shelf—a shallow, submerged region bordering the continents—was a common feature of continental margins around the world. Once imagined as a flat, vast plain, the topography of the deep ocean floor now emerged as a rugged region populated by ridges, mounts, plains, gorges, and trenches.[20]

UNDERWATER CANYONS: SHEPARD'S AND BOURCART'S GEOLOGIES CONVERGE

This was the state of echo-sounding surveys when, in France, Jacques Bourcart understood that those devices could provide him a glimpse of the Mediterranean's submerged geology. Yet it was not simply the use of echo-sounders for research that caught his attention, but some of the scientific results that American geologists were publishing. In particular, the studies of American geologist Francis P. Shepard on underwater canyons (steep gorges) strongly stimulated Bourcart's interest.

Francis P. Shepard (1897–1985) was among the first geologists to dive into the study of shallow-water sediments, supported more by enthusiasm than research means. After completing a PhD in structural geology at the University of Illinois in 1922, he steered his career toward the oceans. From an early age, Shepard became familiar with the ocean, as his father was a passionate sailor in his free time as well as the president of the Shepard Steamship Line and Shepard-Morse Lumber companies in his professional life. During childhood, he spent summer vacations sailing the New England coast with his family on their yacht—which he didn't particularly enjoy due to seasickness.[21] Nevertheless, the summer after graduating, Shepard took his father's forty-foot yacht to do geological work at sea, meticulously sampling seabed sediments off New England shores. This fieldwork became the seed of his growing interest in studying the role of sea level changes in the evolution of the continental shelf—a topic similar to Jacques Bourcart's research on marine regressions in the Mediterranean.

Shepard's specialization in marine sediments came at a timely moment. After the First World War, federal organizations in the US began to support a new model of oceanographic research. Military devices like echo-sounders, which enabled experts to *see* below the mass of water, prompted a sort of techno-optimism in which the seafloor seemed, for the first time, accessible.[22] The University of Illinois awarded Shepard a grant to obtain nautical charts from all over the world, and it supported his findings with the help of researchers at the newly inaugurated Woods Hole Oceanographic Institution.[23] Meanwhile, his research caught the attention of the US Coastal and Geodetic Survey, responsible for mapping the undersea territory. The institution grew interested in using acoustic devices to chart submarine regions annexed to the US, but its experts did not possess the scientific expertise needed to interpret the sounding data that was being acquired. Especially problematic was data recovered off the New England coast, where the Survey had discovered numerous submarine canyons: steep gorges, excavated on the edges of the American continental shelf. Captain Patton, from the Survey, proposed Shepard to collaborate in future campaigns, an offer that was enthusiastically accepted.[24] That same year, Shepard embarked onboard vessels of the US Coast and Geodetic Survey. Equipped with echo-sounders, they acquired large amounts of data from the Atlantic that became fully available for Shepard's research. Those surveys demonstrated that, far from being exceptional, submarine canyons existed along the American continental shelf from Boston to southern North Carolina. The existence of submarine canyons greatly intrigued Shepard, a fascination that intensified in 1934, when he moved his investigations to the California coastline. There he discovered equivalent formations cutting across the continental shelf.[25] At that time, Shepard and his team—comprising geologists Robert S. Dietz and Kenneth O. Emery—had recently joined the Scripps Institution of Oceanography in La Jolla, San Diego (fig. 2). The institution, then a modest marine station, had just a dozen permanent oceanographers, a few small boats, and unreliable sources of funding.[26] From that spot, Shepard would devote a large part of his career to deepening the understanding of the genesis of submarine canyons: Were they excavated by underwater currents? Were they cracks resulting from sudden catastrophic events, such as tsunamis? Or had they been eroded by ancient, once above-ground rivers?[27]

Shepard's studies on submarine canyons, widely disseminated in scientific articles and congresses, immediately aroused Bourcart's interest. They reminded him that, in the late nineteenth century, the French geologist Georges Pruvôt had identified similar formations in the French

Figure 2. American marine geologist Francis P. Shepard, photographed in his office at the Scripps Institution of Oceanography (La Jolla, California) in 1947. Shepard is known as "the father of marine geology," pioneering the field with his studies in underwater canyons. In 1948 he authored the book *Submarine Geology*, which became the benchmark in the field for scientists and the offshore oil industry alike. Source: Special Collections and Archives, UC San Diego, La Jolla. Reprinted with permission of UCSD.

Mediterranean. Pruvôt had sounded and charted the depths along Roussillon's coastline using sounding lines when he realized that the continental shelf had deep submerged gorges. He baptized them "rechs"—a Catalan term referring to channels eroded by rainwater streams.[28] If those rechs were equivalent to Californian submarine canyons, and if they had been created by the coursing of ancient rivers, Bourcart would have strong evidence supporting a massive retreat of the Mediterranean Sea. During the event, mainland rivers would have had to traverse the now-emerged continental shelf to reach the sea, thus eroding steep valleys—now become submarine canyons.

As Shepard's hypothesis around submarine canyons spread internationally through scientific publications, Bourcart began to exhort the French geological community to take a more serious look at the seafloor. To efficiently uncover the Mediterranean's geological history, geologists needed to complement evidence from mainland geomorphology with data from the seafloor.[29] The outbreak of World War II forced Bourcart to postpone plans to pursue the seafloor's geological study in the short term. Unexpectedly, far from being an obstacle, the war in fact eventually enabled him to move his investigations to sea. After joining the French Resistance forces in Paris, Bourcart gained a military reputation for supporting Allied missions against the Nazi-collaborationist Vichy regime, which occupied France between June 1940 and August 1944. At his geology laboratory at the University of Paris, Bourcart compiled cartographical maps to suggest suitable landing spots for the Allied forces, offered geographical training for young recruits, and conducted fieldwork for the strategic defense of Brittany and the Mediterranean coastlines under the pretext of conducting scientific activities.[30] He came to be known by his military superiors as a man of "peculiar intelligence, a scientist of great value with unusual knowledge, a glorious military past, and brilliant clandestine activity," valued not only for his priceless scientific results but also for his courage and tenacity.[31] In one of his secret geographical missions at the Franco-German border, Bourcart was detained by the Gestapo. He managed to escape, liberating nine French prisoners at the same time.[32]

These activities brought him renown within the French Navy which, at the war's end, appointed him as military advisor to the newly created Committee of Oceanography and Coastal Studies (COEC). Comprised of military representatives, the COEC, under the Navy's Hydrographic Service, aimed to promote and coordinate oceanographic studies in French waters. This unique position granted Bourcart the opportunity to shift his geological investigations from the mainland to offshore by providing expert assistance in underwater topography for military purposes.

The creation of the COEC in 1945 was an opportune moment to start marine geological missions: many leading nations were turning to the seafloor not only as a space worthy of military control and surveillance but also as a site holding the promise of future unlimited riches. In this framework, the seafloor began to emerge as a new territory to secure, tame, and explore.

NEW COMMERCIAL HORIZONS: THE POSTWAR TRANSFORMATION OF THE SEAFLOOR

Traditionally, the deep seafloor—beneath the seabed's surface—had been politically perceived as a place unconnected to human activities. Regions beyond the coastline were mainly regarded as surfaces to connect distant regions through cables, or grounds to fulfill the curiosity of scientists and explorers. The seafloor was conceived as a liminal space that could not provide any economic return to coastal countries. At the end of World War II, this perception was about to radically change. International craving for hydrocarbons drew attention to the seabed and its potential natural resources, giving impetus to the development of marine geosciences. Such a movement originated in the US, where marine geology began to receive wealthy public and private support and thus to thrive.

It all started on September 28, 1945. Just a few weeks after World War II officially ended, American President Harry S. Truman declared that natural resources on the continental shelf surrounding the US, belonged to the US.[33] Such a bold statement was framed on a growing thirst for oil among contending nations, urgent to construct extractive industries and rebuild their economies. As a reference, from 1936 to 1950 global oil production exploded from barely 5 million barrels per day to more than 10 million.[34] That President Truman set the future of American energy supplies beneath the ocean was not by chance. The American oil industry had started to spot shallow-water regions as promising for hydrocarbon production two decades earlier, when the swamps and marshes in southern Louisiana turned into productive grounds due to a mix of "geological good fortune and corrupt deals with Louisiana officials," as historian Tyler Priest phrased it.[35] The region was covered by easy-to-identify salt domes, geological structures prone to harbor hydrocarbon deposits. Through a corrupt state system, oil companies obtained extremely cheap leases on public lands.[36] Profitable oil and gas fields found in shallow, inner waters prompted a subsequent interest in the adjacent offshore region: the open waters of the Gulf of Mexico. Its ideal environmental conditions—a wide and shallow continental shelf, with mild weather and barely any waves—transformed the area into an excellent testing ground for the emerging offshore oil industry. In 1934, the oil firm Texaco began to test offshore drilling devices for coastal shallow waters and soft bottoms. One year later, Pure Oil and Superior Oil installed the first productive oil rig two kilometers off the coastline.[37] The shallow seafloor of the US continental shelf began then to be regarded as a new region for exploiting hydrocarbons and minerals, a safe source close to national territory, including refining and consumption areas.[38] In this context, toward the end of World War II, the US Presidential Office set focus on controlling potential submerged resources, eager to speed up the country's economic recovery.[39] The Truman Proclamation benefited American oil firms as well as the national government, since it both ensured US operators that the government was backing their activities and moved the power to issue leases from the hands of state governments to the federal government.[40]

The Truman Proclamation marked a milestone in the history of the seafloor, as it transformed this region into a place suitable for supplying natural resources and, thus, fueling human activities. In that moment, the seafloor began to materialize in the popular consciousness as a space, a piece of land underwater that needed to be explored, tamed, and—perhaps most important—legislated. At the time the Truman Proclamation was issued, no international jurisdiction over marine territories existed; only customary laws and unilateral proclamations ruled on the extension of national seawaters. The issue of the ownership of the seas and oceans, and their natural resources, traced back to the canons of the Roman law, which considered the oceans as *res communis*, that is, belonging to everyone, so no one could appropriate them.[41] Yet this notion evolved toward opposite considerations as independent states emerged and global empires were born. As the Spanish and Portuguese Empires reached the American coastline, the oceans came to be considered as *res nullius*: belonging to no one, and therefore open to all claims. In signing the Tordesillas Treaty in 1494, the two imperial powers divided the ocean space into two exclusive spheres of influence (not on territorial *possessions*), where the Spanish and Portuguese Empires held exclusive control over ocean routes to the east and to the west, respectively.[42] Once again, this principle was contested in the following decades with the emergence of British and Dutch naval powers, which desired to sail freely across sea routes that the Spanish and Portuguese considered their own. In 1608, the Dutch jurist Hugo Grotius argued for the freedom of the seas and their liberation from any national control in the pamphlet *Mare Liberum*.[43] This text became the implicitly accepted doctrine that ruled the oceans for more than three hundred years, although by the eighteenth century coastal nations began to consider their territorial waters to reach "as far as a cannon fired from land can go," or around three miles.[44]

The sinuous evolution of the ocean's legal framework signals the ambiguity of the ocean as a territory. There are no physical landmarks over the vast, blue surface, upon which frontiers could be firmly established. The high seas were challenging to traverse and even more difficult to efficiently control. As the condensed chronology above indicates, the oceans' rule was imposed by those nations who could establish global maritime routes and operate merchant and naval fleets through them. In other words, the high seas' territory, its frontiers, and its uses, were unilaterally delineated by the global empires that, at the same time, ensured compliance. While the large empires conflicted over governing the high seas, for smaller coastal nations more important was to secure control over shallower waters, where their supplies of natural resources were found. Ceylon, for instance, had in 1811 declared national jurisdiction over pearl fisheries, specifying in 1925 that its control extended up to a hundred fathoms depth. Panama had also declared control over fisheries up to 120 miles from its coastline, while Tunisia extended its national control up to 17 miles into the sea to control sponge fisheries. In the UK, the Cornwall Submarines Act of 1858 claimed the right to build tunnels into the seafloor beyond the three-mile limit of the territorial sea, while Australia, Japan, Chile, and Canada undertook coal mining activities in shallow waters.[45]

Against this backdrop, the Truman Proclamation was not unprecedented for being a unilateral assertion of (underwater) territorial governance, but for setting the seafloor and its *potential* oil deposits in the international spotlight. Unlike previous proclamations, this one asserted that technological know-how was already available to transform underwater mineral resources into potential key assets for the national economy. It was also a conceptual breakthrough, since the meaning of the continental shelf was transformed from a geographical feature to an extension of a country's landmass and thus "naturally appurtenant to it."[46] Truman claims were not directly opposed by foreign countries. Quite on the contrary, in the following five years, about thirty countries around the world proclaimed similar jurisdictions.[47] This absence of public opposition can be explained by the fact that, after six years of war, nations were too weak economically, too concerned about quick economic recovery, or too reliant on the US to oppose its new doctrine. For its part, the US dodged opposition by specifying that it would grant drilling rights to foreign governments to explore its coastal waters.[48] Coastal nations understood that, by legitimizing the American movement, they could claim the same rights over their own continental shelves—which, perhaps, might assure them a future source of wealth.

Once the American government had declared the continental shelf as part of the US national territory, US-based oil companies therefore felt under its rule and protection. They began intensive efforts to move their ventures from coastal areas to deeper waters. From 1945 to 1950, forty oil companies invested a hundred million dollars in leases and another hundred million for marine equipment and platforms.[49] The oil firm Shell "pushed aggressively offshore," as its vice president Bob Nanz asserted, by strategically focusing on technological innovation to minimize costs and risks.[50] In 1946, Union Oil struck an alliance with the Continental Oil Company to share costs in a joint project to conduct a systematic survey of the seafloor, hiring geologists and engineers for that purpose. Shell and the Superior Oil Company later joined the team, becoming the CUSS Group.[51] One year later, Brown and Root Marine, an American company devoted to developing offshore platforms and marine pipelines, built the first offshore platform beyond sight of the shore in the Gulf of Mexico, almost twenty kilometers off Louisiana's coast.[52] By 1949, oil and gas companies were crammed onto the Gulf of Mexico's continental shelf, having drilled forty-four oil wells and identifying eleven hydrocarbon deposits in waters shallower than ten meters depth.[53]

The offshore presented a daunting challenge though: the oil industry knew nothing about the marine environment, nor did it possess any proven methods for operating at sea. Before 1945, oil companies that ventured into shallow waters, such as lakes or marshes, had relied on coastal shipyards and naval architects to obtain data and design infrastructures. But the ocean opened new questions and presented new dangers: What sort of waves would they find? How high would they get in a severe storm? How could the piles of platforms be safely fixed? Under which meteorological conditions was it safe to drill? How similar was the seafloor structure to the mainland? Were there the same chances of finding oil and gas? To venture into the oceans, the oil industry required new expertise that could only be obtained from scientific institutions. To recruit academic experts, the technical-trade organization Offshore Operators Committee was set in place in 1948. Representing thirty-two American oil firms, the group aimed at fostering cooperation to confront the complexities of offshore operations by sharing data and activities.[54] Besides deciding that the Gulf of Mexico would be the first area to be thoroughly surveyed, the American Petroleum Institute (API) took responsibility for funding a general, unclassified research program to be conducted at universities and research centers.[55]

The Scripps Institution of Oceanography, Shepard's base of operations and an emergent hub of marine geologists, quickly caught the oil industry's attention. Before and during the war, its experts had conducted military-oriented studies whose outcomes could be also applied to offshore oil operations. Shepard, in particular, had developed coastal studies related to sand dynamics, wave formation, and forecasting to assess landing operations at the University of California War Department. This experience added to his pioneering works on sedimentology, which more or less explicitly encouraged the incipient offshore oil industry. In his recent book *Submarine Geology* (1948), for example, Shepard optimistically affirmed the "immense possibilities of finding oil" in the Gulf of Mexico—and the book soon became a touchstone of offshore oil exploration.[56] That Shepard was at Scripps attracted industry representatives to the institution. They were seeking expert advice and new recruits among PhD researchers while offering to subsidize scientific studies with industrial applications. Shepard was quick in seizing the opportunity to obtain large funds from API. He proposed a research project on sedimentation in the Gulf Coast continental shelf, covering an area of great interest to major oil companies.[57]

From 1951 to 1957, Shepard and his team obtained from API a grant of 100,000 US dollars per year (about 1,080,000 USD today) for their studies. Through the industry's generous support, Shepard had access to unparalleled logistic and technological resources for his research: he used a low-flying plane from Gulf Oil Company to get a seagull's-eye view of the gulf, accessed geophysical profiles of the Mississippi Delta acquired by oil companies, organized missions to sample sediments across the region, met and collaborated with petroleum geologists, and further trained himself in sedimentation.[58] Although he temporarily had to leave his studies on submarine canyons aside, the oil industry provided him with data and borehole samples—materials otherwise inaccessible at Scripps—that offered crucial information about the canyons' formation. Seismic profiles evidenced that more submarine canyons were buried below recent layers of sediment, meaning that, after fluvial erosion and marine reflooding, canyons were filled by the accumulation of recent sediments until their morphology disappeared underneath them.

In addition to supporting Shepard, the American oil industry contributed to the institutionalization of oceanography at universities. Oil companies were increasingly dependent upon academic institutions, such as Scripps or Louisiana University, for their research expertise, but they were far away from the Gulf Coast. In 1949 Claude ZoBell, a marine microbiologist at Scripps, recommended that Texas A&M University create an oceanographic department, a suggestion implemented the following year. The university was located in Texas, only about 150 miles (250 kilometers) from the Gulf of Mexico's coastline. In 1951, the new department launched a major study of the Gulf of Mexico, pursued relevant expertise by institutionalizing training courses, and addressed most of its projects to commercial inquiries. In short, during the fifties and triggered by the oil industry's economic support, marine geology thrived in the US. Industry and science established cooperative relationships, significant funds supported marine geological studies, and new, specialized departments were instituted.

FRANCE'S UNDERSEA PATH AFTER THE WAR

In the US, political focus on the still unknown resources of the seabed lent momentum to the marine geosciences, since potential oil and mineral deposits could only be identified through the improvement of marine geological knowledge and techniques. Shepard's postwar investigations at sea exemplify the new industry-academic alliances that, together with military-academic alliances, led to the formation of marine geosciences.[59] In other words, in the US, a combination of money, oil, and military interest prompted the seafloor's integration into both the geopolitical and the scientific worldviews. The Truman Proclamation marked not only a new vision of the seafloor but also the emergence of new industries and novel research disciplines, while also predicting new international frictions. But did this new imaginary penetrate as deeply in other coastal nations? In France, a formerly powerful global empire now devastated after the war, something different happened. The reason lies in the different geopolitical settings of the two countries and the kind of territories upon which each committed to securing hydrocarbon supplies.

At the end of World War II, France adopted a different approach than the US regarding its control of energy supplies. Given France's rule over North African territories, the nation's future energy security was considered to be found there, not in the oceans. Hydrocarbon and mineral exploration intensified in Tunisia, Morocco, Algeria, and across sub-Saharan regions (as the next chapter details).[60] At the same time, the hexagon's continental shelf remained a region of paramount strategic and military interest—which was equally true for the US and other countries. After the war, French naval forces launched intensive efforts to map, monitor, and control submarine regions next to French territories, both in Europe and abroad. This research included physical oceanography, investigation of the ocean's water mass, marine geology, mapping the seabed's topography, and studying its upper composition.

Coincidentally, the priorities of France's naval forces smoothly converged with the geological interests of Jacques Bourcart, newly appointed advisor in coastal geology to the French Navy for his esteemed collaboration during wartime. The study of submarine canyons in the continental shelf was of utmost military importance, since these gorges could become secret corridors for enemy submarines, enabling them to approach the French coast closely without being detected. The issue was particularly delicate on the French Mediterranean continental shelf, because in the early 1930s secret bathymetric soundings had shown the existence of a deep canyon that led directly to Toulon Bay, location of the main naval base of the Hydrographic Service. During the war, bathymetrical charts remained unfinished and classified top secret, and at war's end the submarine canyon of Toulon remained poorly understood.[61]

Military ambitions to develop deep knowledge about submarine canyons proved a serendipitous opportunity for Bourcart, who used them to transfer his long-dreamed investigations of Mediterranean basin history to the offshore. Due to his position at COEC (the Committee of Oceanography and Coastal Studies), the Navy's Hydrographic Service invited

Bourcart to join, for several weeks each year, the military hydrographic missions that the Office of Oceanographical Studies periodically launched from Toulon.[62] Beginning in summer 1946, he was commissioned to identify and exhaustively describe the submarine canyons that crossed the French Mediterranean continental shelf, enjoying unprecedented access to the navy's research facilities—vessels, echo-sounders, mariners, and classified bathymetric data.

The Aviso *Ingénieur Elie-Monnier* became the main vessel from which Bourcart worked, and its commander was none other than young lieutenant Jacques-Yves Cousteau.[63] Cousteau had started his underwater diving experiments in 1943 while in the French Navy, producing successful prototypes of singular diving technologies—notably the *Aqua-Lung*, the first commercial open-circuit, self-contained underwater breathing apparatus ("scuba" diving, as it came to be widely known). His inventiveness and determination to move human activities underwater caught the attention of the navy's general staff, including Admiral André Lemonnier, who created the Group of Submarine Research so that Cousteau and his team could continue to develop diving devices with the navy's support after the war. The frequent offshore collaboration forged between Bourcart and Cousteau onboard the *Ingénieur Elie-Monnier* symbolized the early entanglement of traditional geology with the development of technological innovations that, in the following decades, would lead to an unprecedented comprehension of the seafloor. Bourcart led a cohort of enthusiastic geologists who continued the research lines of their mentor in the offshore, while Cousteau and his team became leading promoters of cutting-edge technological devices that would give scientists access to the depths. But all this will unfold in the upcoming chapters.

To support himself in analyzing data gathered during offshore missions, Bourcart recruited a young team of recently graduated geologists from among his students at the University of Paris. The geochemist—and only woman in the group—Claude Lalou was responsible for studying the chemistry of sediments and of drawing bathymetric charts, while geologists Maurice Gennesseaux, François Ottmann, and Eloi Klimek worked together on analyzing the recovered sediments.[64] The team sailed along the French coastline, exploring the continental shelf of the Gulf of Lion, the Côte d'Azur, and even around Corsica. Identifying submarine canyons was not an easy task, as the team relied on echo-sounders which were not specifically designed to sound deep, irregular, regions. These devices performed well when sounding flat areas, since acoustic signals bounced over a single plane. But sounding steep canyons complicated the interpretation of data, because sound waves bounced in multiple planes (the canyon slopes, its bottom, the seabed) and were all registered by the sonar's hydrophone. To draw the profile of a submarine canyon, then, Bourcart had to carefully select and sort the numerical data that indicated the canyon's topographical profile—a task for which he believed he possessed "a special intuition."[65] In other words: Technological limitations in data acquisition and processing made of data interpretation a manual and cumbersome task whose results were prone to inaccuracies.

Sounding data were complemented with sediment samples from the uppermost layers of the seabed, recovered by dredging or coring. Dredges, basketlike instruments, were pulled over the seabed's surface to recover large amounts of sediments. With these, geologists could study the nature and composition of the seabed, although sediments from different layers would mix in the dredge, so that dredging was not useful for studying the stratigraphic record. For that purpose, marine geologists used corers, cylindrical instruments that, after being nailed into the soft, upper sediment, recovered undisturbed cylinders of sands and muds. To such technological limitations was added the problem of determining the vessel's location accurately. As vessels were not equipped with any radiolocation system, the team of geologists had to situate their sampling position—and thus, that of the data gathered—with sextants, which decreased the accuracy of the submarine charts produced.

After delivering detailed topographical information of strategic submarine canyons, such as the rech Lacaze-Duthiers, a deep submarine canyon facing the frontier between France and Spain, the French Navy commissioned Bourcart and his team to draw a detailed chart of the French Mediterranean morphology up to 2,000 meters deep, by locating underwater canyons and detailing their depth, the nature of their sediments, their origin and ending points. Starting at the Côte d'Azur, between Marseille and Toulon, they proceeded charting southward to the Spanish border, as well as around the island of Corsica, with the support of the Hydrographic Service (fig. 3).[66]

The research focus of these expeditions converged with the pioneering studies that Shepard had been conducting, for the past two decades, on the American continental shelf. Bourcart's hypothesis on the formation of the Mediterranean submarine canyons coincided with those of his American counterpart: Submarine canyons probably had been excavated by ancient rivers, being thus key evidence of massive regressions in ancient oceanic settings. Proof of their fluvial origin were the canyon's features, now glimpsed through topographic profiles: the distinctive V-shape of the walls, similar to that of any other mainland river, the winding path of the canyon, fed by small gorges on its sides, and the walls' composition of hard rocks. Bourcart stressed that the rivers had eroded these valleys during a rapid process, like a torrential water stream, because at the canyons' feet, he recovered gross fluvial pebbles up to 3 centimeters in diameter. Only a powerful river flow could have moved them.[67] By correlating these findings with stratigraphic

Figure 3. Topographic chart of the continental shelf along the French Mediterranean coast, depicting the underwater relief with 100-meter contour intervals. Jacques Bourcart directed its compilation based on surveys conducted by the French Navy's Hydrographic Service. Source: Jacques Bourcart, ed., *Congrès géologique international: Comptes rendus de la dix-neuvième session, Alger, 1952*, section 4, IV (1953). Muséum National d'Histoire Naturelle, Pr 5350–19–1952.

evidence from the mainland, Bourcart dated the canyons' excavation to a recent geological period (between five and seven million years ago, during the late Miocene), venturing that the connection between the Atlantic Ocean and the Mediterranean Sea had been cut off. As no other water source had poured into the Mediterranean to resupply water lost by evaporation, sea level would have massively descended by about 2,000 meters from its initial level, transforming the basin into a mosaic of brackish and isolated lagoons.[68] As a key piece of evidence, Bourcart pointed to Toulon's submarine canyon foot: it ended at 2,200 meters deep in the form of a river delta, which indicated that, in the late Miocene, the river mouth was located there.[69] The Atlantic connection was reestablished in a later phase through the region we now know as the Gibraltar Strait, refilling the basin with seawater and transforming it into the Mediterranean Sea we now know. In other words, the Mediterranean basin could have once become a desert.

INTERSECTING PATHS IN THE MEDITERRANEAN GEOLOGY

If until this point the convergence of Bourcart and Shepard had been essentially intellectual, by the late fifties it became a personal friendship. The French Navy's economic and logistical support enabled Bourcart to reach fascinating conclusions on the origins of submarine canyons, while the

Mediterranean seabed began to take shape through his offshore cruises. However, these studies were not enough to investigate below the seabed's surface. With the navy's support, Bourcart could only know the seafloor's external shape, but not its composition below the surface of the seafloor. Bourcart's team needed the expertise and research technologies that were being developed in countries like the US, the UK, or Sweden, supported by the offshore oil industry or by governmental actors. To penetrate deeper into the Mediterranean's history, thus, it was not only necessary to get his feet wet, but to rely on international collaboration as well.

With this aim in mind, Bourcart first contacted Francis Shepard in 1950, who was about to start his API-supported research on the Gulf Coast's sedimentation. They maintained a frequent correspondence and forged a strong, personal friendship strengthened by Shepard's visits to France. Through Bourcart, Shepard was introduced to the French community of geologists, spreading his theories and knowledge about marine geomorphology. French geologists considered him "the man of submarine canyons," an elegant, sporty, and gentle teacher whose enthusiasm persuaded a young generation to follow him into the depths.[70] Shepard would also fondly remember his trips to France escorted by Bourcart who, in addition to touring him around the University of Paris and the geography of France, would treat him with champagne, wine, and liquor at the slightest opportunity.

In 1955, Shepard and his wife Elizabeth traveled to France, where he spent two summer months exploring the Mediterranean submarine canyons with Bourcart. The two couples traveled by car—Mme Bourcart driving, so the men could indulge in some glasses of wine for lunch—to the summer village of Villefranche-sur-Mer, where the first laboratory of marine geology had just been established at the Oceanographic Observatory under Bourcart's leadership (fig. 4).[71] Villefranche's marine base was among the oldest in France. Founded in 1885 by naturalist Alexis Koronteff as the Russian Laboratory of Zoology, French and Russian experts shared it until 1933. The Russian presence resulted from the Franco-Russian alliance established in the mid-nineteenth century. After the 1856 Crimean War deprived Russia of access to the Mediterranean Sea, the Kingdom of Sardinia offered the Russian Imperial Navy anchorage in Villefranche as a symbol of their alliance. The marine zoological station sat in the dock buildings where the Russian fleet had been stationed.

The spot was ideal for conducting marine geological experiments: Villefranche-sur-Mer is located at the foot of the rugged maritime Alps, facing a close, calm, and deep harbor (its center easily reaching a depth of 100 meters). Facing south, with a narrow connection to the sea, the site is protected from strong winds and waves, and devoid of turbulence

Figure 4. Marine geologist Jacques Bourcart, photographed at sea during the 1950s. Bourcart was a pioneer of marine geology in France and played a key role in establishing the first marine geology laboratory on the French Mediterranean coastline, east of Nice. Combining his lectures at the University of Paris with research off Villefranche-sur-Mer, Bourcart mentored a first generation of marine geoscientists in France. Photo by Claude Guernet, courtesy of Gilbert Boillot. Reprinted with permission.

from freshwater streams. Perhaps more relevant for offshore missions, Villefranche is only 150 kilometers away from Toulon's naval base, where Bourcart and Shepard joined the *Ingénieur Elie-Monnier* to explore submarine canyons in front of Nice's coastline and around Corsica. The trip gave Shepard the opportunity to make soundings and collect samples while strengthening links with French geoscientists. Both geologists, the American and the French, shared data, ideas, and a hypothesis, extensively discussing the possibility that the Mediterranean submarine canyons were created by fluvial erosion over a region that emerged in a distant past. Shepard's experience relied on more than twenty years of pioneering research along the California coastline, the shallow continental shelf in the Gulf of Mexico, and the regions he first studied off New England. For those investigations, he had received economic and logistical support from the US

Geological Survey, the wartime naval forces, and the collective American oil industry. Bourcart, for his part, built his firsthand understanding of the Mediterranean seafloor from his extensive experience studying mainland geology. The French naval forces supported his marine experience by furnishing him resources and the opportunity to embark on some coastal expeditions. The research pathways of the two scientists illustrate the different trajectories of marine geosciences' development on opposite sides of the Atlantic Ocean.

This sort of international collaboration to unveil the Mediterranean's history did not extend beyond sporadic and bilateral personal collaborations— even though Bourcart tried hard to convince other foreign experts, in addition to Shepard, to join explorations of the Mediterranean seafloor. In summer 1955, Bourcart organized the International Colloquium of Submarine Geology in Paris. It was the first of its kind in France. He invited internationally renowned oceanographers, specialists who could potentially contribute to Bourcart's scientific goals with logistical resources and expertise. For example, the Dutch geologist Philip Kuenen, director of the Geological Institute of Groningen, had in 1950 published the foundational book *Marine Geology*. The Swedish oceanographer Hans Pettersson had recently led the circumnavigation expedition of the *Albatross*, testing cutting-edge devices to explore the seafloor, like echo-sounders that could obtain continuous depth profiles in up to seven kilometers' depth, and the "piston core sampler." Better known as the Kullenberg corer, this hydraulic coring instrument was capable of recovering twenty-meter-long cores from the seafloor.[72] With these, Pettersson and his team demonstrated the thickness and age of soft sediments across the oceans, providing the basis to accurately determine the time, duration, causes, and consequences of the Ice Age.[73] Shepard, of course, could bring in the American community of marine geoscientists and the sedimentological expertise that was being developed with the oil industry's support.

Bourcart had hoped to persuade the attendees of this renowned colloquium to organize a large-scale marine geological survey across the Mediterranean basin, bringing together French, American, Swedish, and Danish experts.[74] Despite optimistic discussions on the anticipated expedition's logistics and calendar, the joint campaign never materialized. Bourcart had been the one commissioned to provide a research ship, but there were scarce research resources in France for academic investigations at sea. Beyond its limited access to naval vessels, the French oceanographic community did not possess a large oceanographic ship that could be equipped with devices like the piston corer, nor was it prepared to host a large scientific party. Lacking the main elements—funding and ships—to conduct large oceanographic missions, the development of French international collaborations was handicapped in the early stages of marine geology.

A LIFETIME SEARCH FOR THE SEAFLOOR'S GEOLOGICAL PAST

Bourcart's experiences show how studying the Mediterranean seafloor changed throughout the first half of the twentieth century. If in his youth *seeing* the seabed's mounts, canyons, and plains was a future Bourcart could only dream of, in 1958 visualizing the underwater geography had become the basis of his research. Just before retiring in 1961, he summarized these transformations by highlighting that, despite a "long-established lack of funding," French marine geology now enjoyed an oceanographic base, access to vessels, research tools, and an emerging and enthusiastic workforce trained to study the Mediterranean in the offshore.[75]

Settled in the newly established laboratory of marine geology at Villefranche-sur-Mer, Bourcart and his students spent summer mornings doing routine sampling onboard small fishing boats adapted to marine geological research. For wider surveys Captain Jacques Cousteau, just appointed director of the Oceanographic Museum of Monaco, invited them onboard the *Calypso*. There, they could use echo-sounders and sampling devices—dredges, to pull up superficial gravels, sands, and muds, and gravity corers, to retrieve cores up to one meter in length. The lab analyses of recovered fine sands, mixed with planktonic foraminifera and wood scraps, fluvial gravels, and mud samples led the team to reach conclusions far beyond the continental shelf, about the catastrophic events that could have drastically transformed the Mediterranean region millions of years ago.

Bourcart taught his students that for conducting marine geology, a small boat, a winch, and willingness were more than enough. Yet despite his optimism, Bourcart's research style proved insufficient to *see* what lay beneath the first meters of soft sediments: What geological formations constituted the seafloor? And what could they explain about past dynamics of the oceans and continents? These questions could only be answered by applying marine geophysical techniques, which provided numerical information about the seafloor's deep composition and structure. Operating them required specific technologies and well-equipped vessels, large amounts of funding to spend time at sea, and a different set of expertise stemming from physics. Foreign researchers were starting to conduct this sort of campaign, either boosted by the oil industry or framed in international cooperative programs of ocean exploration.

By contrast, in France, without the search for oil as a driving force, marine geological studies were constrained to the seabed's surface. The military exploration of France's colonial territories had first enabled Bourcart to acquire an overall perspective on the Mediterranean's geology and history. Naval underwater priorities after World War II converged with Bourcart's scientific interests in studying submarine canyons and charting the seabed's topography. However, the French Navy's approach to seafloor exploration constrained research in other directions, impeding the study of deep regions, the organization of cruises around the world, and the development of marine geophysical techniques to study the deeper stratigraphy of the seafloor. In other words, the navy's support for Bourcart's activities was insufficient to organize larger, systematic studies, to acquire more advanced research technologies, or to coordinate a critical mass of disciples with relevant expertise in marine geology, which would in turn enable them to move their research lines to an international level.

These limitations inspired deep concern within the French scientific community, which worried about losing the scientific prestige associated with being first to reach new conclusions about the Mediterranean seafloor, the region they had traditionally studied.[76] John B. Hersey, American oceanographer at the Woods Hole Oceanographic Institution, began exploring the Mediterranean seafloor in summer 1957, stimulated by the geological potential of the region to advance knowledge about crustal tectonics. He used echo-sounders to define the bathymetry and morphology of the deepest regions, and employed new, pioneering marine geophysical techniques, including refraction seismics to understand crustal features and reflection seismics to study sedimentary sequences accumulated over the basement. Maurice Ewing, from the Lamont Geological Observatory, launched Lamont's first campaigns across the Mediterranean basin on the vessel *Vema* in summer 1956, returning the next year equipped with cutting-edge seismic refraction devices.[77] English researchers from the University of Cambridge and the UK's National Institute of Oceanography were also venturing into the western Mediterranean equipped with marine geophysical techniques to sound its deepest structures.[78] From the perspective of Bourcart and his team, pushing forward with marine geophysical research was no longer only a matter of scientific curiosity, but had become a pressing issue of avoiding a "regrettable delay," as they phrased it, for France to pioneer seafloor exploration.[79] In order to attract funding to marine research, Bourcart began to forge a public discourse that became a mainstay in academic marine geosciences: a call to the "Public Powers" urging them invest in seafloor exploration. His arguments were both nationalistic and economic, namely to increase the nation's international prestige via oceanographic capabilities and, subsequently, to benefit from the economic possibilities that this new territory might offer.[80]

Geopolitical changes in the first half of the twentieth century had a significant impact on how the seafloor was understood and explored: It progressively began to emerge as a space that needed to be controlled and regulated by political, military, and commercial powers. While this theme will recur throughout the following chapters, Jacques Bourcart's career has offered a unique lens through which to explore the beginning of this transformation.

Bourcart's life and research unfolded during a period of profound transformation—both in France's global position and in the geosciences. The French Empire's extent and the national desire for territorial control led Bourcart to start his career mapping the geology around the Mediterranean using traditional methods—"boots and a hammer." As the military strategic importance of the seabed grew, Bourcart was able to shift his research toward the sea. However, while the imperial and military visions of France guided and supported Bourcart's geological research in certain directions, they also constrained it in others. These limitations are well illustrated by comparing Bourcart's work with that of his American counterpart and colleague Francis Shepard. The two men converged in studying submarine canyons, but while Shepard is widely recognized as the "father of marine geology," Bourcart remains known primarily within a small circle of French geoscientists. This does not suggest that Bourcart could have necessarily achieved global fame with more economic support—scientific prestige is tied to an individual's contributions, international influence, and long-term impact on the field. However, this contrast highlights the influence of divergent national visions on seafloor exploration: While the US government and industries heavily invested in offshore development—and thus, in marine geosciences—the French government, confident in colonial energy resources, did not see the same urgency in exploring the seafloor. Bourcart struggled to find the support that would move his investigations beneath the seafloor's surface.

As noted, Bourcart lived on the verge of a world that was fading. He retired just as the French colonial empire was in decline, global energy demands were rising, and the Cold War was dividing the world. The exploration of the oceans and their seafloor was also entering a new phase, driven by new technological capabilities, international collaborations, and complex institutional networks. Bourcart's life and research style, in many ways, represented the end of an era—for both seafloor geology and France's global influence. While the career of an individual has offered a focused lens on this chapter in history, other units—networks, transdisciplinary teams, and institutions—became the key drivers of seafloor exploration as it transformed into a critical national interest and a promising commercial frontier.

CHAPTER 2

Lagging Behind?
Geopolitics and Geophysics
Underwater

"We must go and see."
MOTTO OF JACQUES COUSTEAU

For world-acclaimed commander and adventurer Jacques Cousteau, humans could only decipher the ocean's secrets by physically traveling to and seeing this mysterious space for themselves. Scuba diving, underwater cameras, manned submersibles, and undersea habitats would surge in the upcoming decades, fascinating scientists and the lay public alike with their possibilities. But in 1951, when Cousteau first articulated his motto, *il faut aller voir*—"we must go and see"—during the *Calypso*'s inaugural expedition to the Red Sea, France's seas could barely be (metaphorically) *seen* by marine geoscientists, who had such limited ability to "go."[1]

Cousteau's invitation for wider audiences to turn their attention toward the oceans can be reinterpreted as a call to foster marine sciences in his home country. As a frequent collaborator of ocean scientists, Cousteau was aware of the limitations this community experienced: scarce political attention, a reduced budget, and a small and scattered community of practitioners. All these features prevented the organization of large-scale oceanographic surveys. When Cousteau commenced his expeditions aboard the *Calypso*, French geologists could only visualize the *surface* of France's seafloor through geological methods, echo soundings, and the construction of models and maps. Marine geophysical technologies, already in use by the most capable oceanographic nations, could open up the seafloor's vertical dimension for France too—if it weren't for financial constraints.

France's commitment to seafloor exploration was shaped by a blend of national and international politics, long-standing ideas about the strategic importance of marine resources, and the new geopolitics of the oceans; it was set against the dominant American influence and France's desire to restore its *grandeur*. This chapter explores how France's (scientific) position in the oceans changed at the end of the fifties and, with it, the possibilities of studying the seafloor's vertical composition. In 1958, President Charles de Gaulle brought a new geopolitical vision to the office, and marine sciences suddenly fit smoothly into the political agenda to strengthen France's international position. The confluence of two external events granted marine sciences this novel dimension: the new era of international collaboration to explore the oceans that the International Geophysical Year (1957–58) inaugurated, and the first UN Convention on the Law of the Sea (1958). Both events revealed the impending significance of national oceanographic capabilities in geopolitical relations while exposing the limitations of France in that aspect. References to "lagging behind"—both economically and in terms of international recognition—became key reasons for promoting seafloor exploration in France. These concerns helped drive the initial growth of marine geosciences, particularly in exploring the vertical dimension of the seas. The western Mediterranean, in particular, developed as a hub for French and American geoscientists and, after the first results of its deep composition surfaced, for oil companies as well.

THE GEOPOLITICS OF INTERNATIONAL SCIENTIFIC COLLABORATION

The International Geophysical Year (IGY) constituted a tipping point in the relation between national prestige and international performance in Earth and marine sciences.[2] From July 1957 to December 1958, research teams from sixty-seven countries undertook synoptic studies to acquire geological, hydrographic, and atmospheric data around the world, which resulted in unprecedented conclusions about global processes.[3] From the perspective of French oceanographers, the IGY exposed the weaknesses of their country in ocean exploration in comparison to the emergent oceanographic powers, especially the US.

The IGY was not the first international collaborative endeavor in ocean exploration; rather, the very nature of the oceans had traditionally invited international cooperation. Beginning in the early twentieth century, the need to operate advanced and exclusive technologies in costly and long campaigns, and to coordinate the collection and interpretation of data, gave ocean sciences an international dimension. Precisely because of the need to negotiate joint ventures between communities of different regions with diverse national interests, ocean sciences have traditionally been laden with political and diplomatic connotations. The International Council for the Exploration of the Sea (ICES), for instance, was established in 1902 to debate practical applications of ocean sciences for fisheries while providing a model for international science collaboration in the aftermath of World

War I, when international tensions peaked.[4] Another example was the Pacific Science Congresses that, throughout the 1920s, fostered exchanges and coordinated efforts between countries bordering the Pacific Ocean.[5] The aftermath of World War II provided a renewed impetus to international collaboration as new research technologies, like sonar systems and geophysical devices, forged avenues toward a new understanding of the marine environment. New communities of oceanographers, trained under the US Navy's mentorship, came to dominate the scene, while bountiful funding supplied by new patrons—national navies, oil and extractive industries—began to flood oceanographic institutions, particularly in the US. At the heyday of international political tensions, cooperative scientific research in the oceans acquired a political undertone: oceanographic missions became a means to build alliances, a mechanism through which national governments could obtain information, and even a way to access the technoscientific capabilities of other nations.[6]

With the traditions of international cooperation well in place, the historic significance of the IGY lies rather in its transformation of that cooperation to larger scales, making it a common practice for exploring the oceans—through programs where scientific and geopolitical interests were interwoven.

The IGY was initially conceived as a scientists' program, in which national governments should not be involved. In the organizers' discourse, the project aimed at being devoid of Cold War politics, so that *basic* scientific research was pursued for a common benefit. However, as historian Jacob Hamblin has emphasized, despite this rhetoric, the IGY was politically laden from its inception. Some countries feared that IGY research in Antarctica would establish precedents for future territorial claims; the US government was reluctant to include Soviet scientists in the IGY global data network; and one of the IGY's oceanographic programs consisted of identifying deep oceanic areas where water was relatively stagnant, so that radioactive waste might be deposited without harmful effects.[7]

For French oceanographers, the IGY provided a unique opportunity to establish cooperative relations with the world pioneers in ocean exploration. But in doing so, the weaknesses of France's capabilities were exposed and felt by its participants. Henri Lacombe's experience at IGY offers an illustrative example. Lacombe, French naval hydrographer and oceanographer, enjoyed the distinction of occupying a new chair of physical oceanography in France, opened at the National Museum of Natural History of Paris in 1955. Lacombe was invited to participate at the IGY Oceanographic Research Working Group by Robert S. Dietz, an oceanographer at the US Scripps Institution of Oceanography.[8] Aiming to make the most of this opportunity, Lacombe gathered experts and the limited research resources available in France to organize a Franco-Spanish survey. The joint team studied seawater exchanges through the Strait of Gibraltar, a topic that, aside from its scientific interest—Lacombe considered the Mediterranean a miniature model to study the global ocean circulation—was of utmost geostrategic importance. For the US, the Strait of Gibraltar was key to controlling the traffic of Soviet submarine fleets that could spill into the Atlantic Ocean from the Black Sea. An accurate knowledge of submarine currents and the exact location of the thermocline was essential to accurately operate sonar systems and thus detect the movement of Soviet submarines.[9] Despite the scientific success of the survey, due to limited governmental support Lacombe experienced handicaps to French participation. He had committed to contribute three French vessels to the international mission; however, he could only bring two because, aside from the Hydrographic Service, no other sector from the public administration offered them a third.[10] In a report addressed to the prime minister's cabinet, Lacombe denounced the limitations he faced, and he stressed that participation turned out to be possible only "thanks to the good will" of French researchers, who had pooled all their available resources to support the project.[11] He warned that France risked a loss of international prestige if the situation was repeated—a warning that became an omen for the near future.

After the IGY's international success, programs and committees to explore the oceans multiplied. As oceanographic capabilities crystallized into symbols of a nation's stance in the world order, French researchers' concerns deepened. Henri Lacombe became France's representative in almost all the international forums, a position that mirrored the relatively small community of practitioners: he appeared at both the Special Committee on Oceanographic Research (SCOR), an international, nongovernmental committee gathering scientific associations from around the world to continue oceanographic programs after the IGY, and at the International Indian Ocean Expedition (IIOE), the global program SCOR conceived.[12] Lacombe sat next to world-renowned oceanographers, like the Danish Anton Brunn, British George Deacon, and American Roger Revelle, at the International Advisory Committee on Marine Sciences (IACOMS), an advisory group for UNESCO on ocean activities. He played the same role at its subsidiary intergovernmental mechanism devoted to education, training, and cooperation in ocean sciences, the Intergovernmental Oceanographic Commission (IOC).[13] Through Lacombe, France was expected to contribute to international efforts with experts, vessels, technologies, and laboratories.

International cooperative projects were also growing oriented toward more than the peaceful exploration of the oceans. In late 1958, American anxieties grew stronger for the Mediterranean basin after a secret report issued by the British navy showed how the Soviets were gaining advantage over them in charting—and controlling—the oceans.[14] To prevent the Soviets from winning further control, NATO commanders decided to invest in developing scientific knowledge of the Mediterranean and North Atlantic, essential in order to enhance military capabilities for underwater surveillance, operations for submarine warfare, weather prediction, marine transport, and the prevention of the effects from radioactive fallout. In September 1959 NATO's Scientific Committee created the Subcommittee on Oceanographic Research (ORC), responsible for organizing a coordinated oceanographic program among NATO member states. As a member of the alliance, France was called to actively contribute research resources to the effort. Henri Lacombe once again represented his country.[15]

These appointments concerned Lacombe, who knew the negligible importance of academic marine sciences in the French government's agenda up to this point. His experience in physical oceanography mirrored the situation of the other branches of science, including marine geology. In France, the limited public investment in university laboratories and marine stations had prevented the development of scientific marine research at a level comparable to that of countries like the US or the UK. The lack of well-equipped oceanographic vessels, suited to survey the high seas, impeded the development of physical oceanography in public laboratories and universities. Lacombe and his team could only pursue studies on underwater currents and tides with the occasional logistical support of the French Navy's Hydrographic Service—a situation similar to the one Bourcart was facing in marine geology. Marine research was dispersed across dozens of small, ill-equipped laboratories and marine stations which, in turn, were mainly used to train undergraduate students in zoology and algology. And the absence of an academic training plan prevented the development of future marine scientists.[16]

With the advent of international cooperative programs and committees, researchers increasingly lamented the poor situation of French marine research, highlighting the country's inability to effectively represent itself. Their strategy to attract governmental funding became a plea to increase national prestige through marine sciences, expressed in different formats and venues. Bourcart, for instance, expressed his concerns in publications for wider audiences. In his handbook for students in marine geology, he highlighted that he could accomplish the research there presented despite his "limited resources," which were much lower than those in the US.[17] In a book about the ocean floor for a more general audience he emphasized that, although France bordered on "the three main oceans," its contribution to knowing the oceans was merely a "symbolic" one and that American, British, and Scandinavian vessels were conducting most of the work.[18] More boldly, Lacombe asserted in a governmental report that the nation's poor capacity to contribute to international oceanographic forums could lead to an embarrassing and humiliating position.[19] Scientific capabilities displayed in international committees and programs should reflect ambitions for France's economic and political place in the world order, yet French oceanography at the moment was far from mirroring any such power. It was thus urgent, his argument continued, to invest in equipping laboratories, building vessels, designing new technological devices, and training a body of experts who could adequately represent the country.[20] Lacombe called for an investment in what he called basic science (research developed at public laboratories, without any direct goal or application), insisting that this was the kind of research pursued in international oceanographic programs, although he knew that results were also applied for military surveillance purposes, such as in NATO's cooperative programs and during the IGY's exploration of the Strait of Gibraltar. These repeated calls to governmental interests aimed to attract the attention and financing of institutions and organizations like the Ministry of Research, the Ministry of Industry, or the army, to promote a national scientific agenda.[21]

Specifying the importance of investing in basic research mattered because France held a long tradition of supporting applied marine research. Fisheries had traditionally been a major economic domain for France, and the overseas territories in the Pacific, Indian, and Atlantic Oceans facilitated the exploitation of fishing grounds far away from the hexagon. In 1918, the country had established a specific institution addressing scientific and technical development to "exploit the riches of the sea": the Scientific and Technical Office of Maritime Fisheries, through which France joined ICES. At that moment, "riches" translated to different kinds of fisheries (like herring, tuna, and oysters), processing industries, and canning industries. In 1953 that office was restructured and renamed the Scientific and Technical Institute of Maritime Fisheries, with a broader mandate that encompassed applied oceanography and marine biology. It possessed its own oceanographic vessel, the *Président Theodore Tissier*, exclusively devoted to studying fish biology, fishery migration, and the cartography of fishing grounds.[22]

The imbalance between the public resources devoted to applied research (fisheries) and those invested in so-called basic investigation (physical oceanography, marine geology, marine biology) distressed Lacombe, who anticipated the geopolitical relevance the oceans would gain in the next years. He argued that an international race to learn about the oceans' strategic and economic potential was about to start, and it would be run via scientific research. In addition to physical oceanography to learn about the mass of water, marine geosciences would play a central role in knowing the seafloor, and marine biology in enhancing the exploitation of living resources.

MARINE SCIENCES TO REGULATE THE OCEANS

Researchers voiced their concerns about the scarce French representation in international forums and their limited budget for marine sciences at a timely moment. In the same months, international negotiations on a new ocean legal framework prompted the French government to quickly enhance scientific exploration, anticipating its significance in shaping the country's position in global ocean affairs.

In France, political urgency for marine research emerged during the first UN Convention on the Law of the Sea (UNCLOS I), inaugurated on February 24, 1958. More than seven hundred delegates from eighty-six countries gathered at the Swiss city of Geneva to reach a legal consensus on the rights of coastal nations over the oceans. The need for the conference had grown from the international turmoil caused by the Truman Proclamation, when numerous coastal nations around the world unilaterally proclaimed their rights over extended marine regions.[23] Prevailing international relations deeply configured the conference's debates. The disputes between the Eastern and the Western blocs suffused the agreements, while a growing North-South fracture began to appear. New nations, driven by anticolonial sentiment, feared that the Law of the Sea would become a new tool to extend colonial power.[24]

The main contentious issues were related to military security, commercial navigation, and fishing rights. To give an example, while the US pushed for a narrow territorial sea (up to three miles) to be able to freely move its forces through strategic points like straits and narrow seas, the USSR advocated for a twelve-mile limit—which, from the US perspective, risked jeopardizing global maritime commerce. National control of fisheries, both in the high seas and over the continental shelf, was also a divisive issue: developing nations pushed for extended rights, fearing that the new legal framework would limit their control over fishing grounds. After two months of intense discussions, the conference adopted four conventions concerning the definition and freedoms on the high seas, the coastal-state jurisdiction over the seabed and its subsoil resources, the territorial sea and its contiguous zone, and coastal-state rights on fisheries.[25]

For the French government, the first two of these conventions triggered concerns about the country's competitive position on the international scientific stage. There was also unease about France's ability to generate the scientific knowledge needed to define its geopolitical stance in the oceans. The high seas were consensually recognized at UNCLOS as "open to all nations, [where] no State may validly purport to subject any part of them to its sovereignty," while acknowledging four international freedoms that ensured global access: navigation, fishing, laying submarine cables and pipelines, and overflight.[26] Marine research, although discussed as a potential fifth freedom, was not included in the final text of the convention due to concerns over sovereignty and national security. International cooperation to explore these regions emerged as a prime strategy for easing diplomatic tensions or strengthening alliances. From the French government's perspective, thus, competent participation in joint campaigns across the high seas could enhance its country's prestige and foster international relations, but in order to do so, France needed to first develop those research capabilities at home.

At the same time, the need to define the breadth of territorial waters and to draft a precise definition of the continental shelf became key drivers for studying France's surrounding oceans in depth. These issues were central to the organization of the conference, prompted by the rapid growth of offshore industries. Their expansion made it essential to precisely define the national boundaries over which coastal nations could exercise jurisdiction.

In particular, the definition of the continental shelf—the area over which resources like oil and gas might be exploited—became a contentious issue. The challenge lay in transforming a geological concept into a legal one. The International Law Commission's preparatory works had proposed a flexible definition, enabling nations to expand their jurisdiction as technological capabilities advanced to access deeper resources:

> The seabed and subsoil of the submarine areas adjacent to the coast but outside the area of the territorial sea, to a depth of 200 meters or, beyond that limit, to where the depth of the superjacent waters admits of the exploitation of the natural resources of the said areas.[27]

France's representative, international jurist André Gros, led a contingent of states that opposed this definition, which they deemed inconsistent, ambiguous, and subjective. According to Gros, the incomplete state of scientific data on the geology, geography, and oceanography of the continental shelf made it difficult to establish a clear and effective definition. He was particularly opposed to using "possible exploitation" as a criterion for delimitation, arguing that it was inherently subjective. He questioned whose capabilities would become the standard: those of the most technically advanced states, or the capacity of each state in relation to its own continental shelf?[28] Such criteria would only exacerbate animosities between ocean powers and less technologically developed countries, since these latter could secure large extensions of the seafloor without being capable of an actual exploitation, using as pretext their *future* capabilities.[29] As a result, France refused to sign the Convention on the Continental Shelf.[30]

Against such anticipated competition for moving resource exploration toward deeper waters, France found it crucial to start generating knowledge about its territorial waters. Only if equipped with that knowledge could the country be ready to accurately define its position in future UN conferences, in which the limits of the national continental shelf would be renegotiated. Recall that French territories extend over four oceans, and that the hexagon needed to be defended along two wide coastlines (the North Atlantic and the Mediterranean). By identifying potential sources of natural resources on its continental shelf, in the hexagon and around its overseas territories, France could bring up new arguments in future UN meetings to propose a maritime frontier that would benefit its current and future ocean industries.

UNCLOS I marked the legal transformation of the seafloor into a political territory, as it defined one of the territory's constituent elements: its boundaries. For coastal governments, now the oceans were divided into internationally agreed regions, ruled by different ownerships. The geostrategic importance of the oceans was no longer only a matter of military security but also an issue of securing potential natural resources. Science and technology became central elements for its exploration, through peaceful and internationally accepted means.

MARINE SCIENCES TO RESTORE FRANCE'S *GRANDEUR*

Peak international interest in exploring the oceans coincided with the election of General Charles de Gaulle to France's presidency in December 1958. In the pursuit of transforming France into a major player in global affairs, his office found it imperative to secure a maritime presence comparable to that of other nations. In this context, marine sciences seamlessly integrated into the new science policy and began to thrive.

Charles de Gaulle of course was already a prominent figure before becoming France's president. During World War II, he led the exiled French government during German occupation and became the head of the provisional government of the French Republic in June 1944, after France's liberation. However, his disagreements with the subsequently elected government led him to retire from politics in 1946. On May 13, 1958, the Algerian coup d'état, fueled by Gaullist activists, pushed France to the brink of a civil war. In response, de Gaulle returned to the political stage, driven by his belief in the need for a stronger government for France and encouraged by his supporters. This sector, including military leaders, intellectuals, and business figures, viewed him as the only leader capable of restoring order and reestablishing France's international prestige. The French National Assembly transferred power from President René Coty to de Gaulle, who was appointed president of the council. He then drafted the constitution that established France's Fifth Republic in September 1958. Two months later, he was inaugurated as its first president.

De Gaulle undertook major political reforms to restore what he called France's *grandeur*—the nation's international prestige that had been lost during World War II. His political agenda aimed at transforming France into a nation more independent of the greatest powers, the US and the USSR, counterbalancing the influence of both in Europe. This stance was captured in numerous geopolitical moves that have historically defined his policies. In 1958, for instance, de Gaulle started pushing for a tripartite leadership between the US, Great Britain, and France in the decision-making about nuclear weapons management—which became a central reason for the cooling of diplomatic relations between France and the US. France's drastic change of position with respect to NATO was similarly representative of President de Gaulle's worldview as well as reflecting the growing importance of the oceans. In 1964, de Gaulle withdrew the French Mediterranean fleet from NATO's command after suspecting that the American and Italian governments were secretly supporting Algerian independence.[31] Disagreements about nuclear deterrence further strained Franco-American relations until March 1966, when de Gaulle withdrew from NATO's Allied Command, announced that France would not join NATO's nuclear planning group, and refused to maintain NATO's headquarters in Paris. In this way, France became the only full member of NATO whose naval forces were not part of the alliance. France's Navy could freely operate in line with its national interests, emerging as a symbol of international power and a tool of coercion. Unlike those of other European member nations, France's fleet could survey beyond European seawaters, around its overseas territories; while under NATO's leadership, French submarines played a role in nuclear deterrence to protect the Atlantic and Mediterranean coastlines.

From de Gaulle's perspective, scientific and technological development were central pillars of the nation's economic independence as well as key assets to increase France's national and international prestige.[32] As historian Gabrielle Hecht has shown in her notable work, technological prowess in nuclear research was central to rebuilding France's national identity and self-perceived glory. Similarly, Walter McDougall has detailed how the French space program sought proactively to increase France's international prestige.[33] The trope of "lagging behind" other countries such as the US, the UK, or Germany acted as a catalyzer to stimulate domains ranging from the technosciences and national economy to domestic industries alike. For the sciences, in particular, it provided justification to promote a centralized, politically driven, scientific policy.[34] Explicit references downgrading scientific development in France, as well as the political fear of "France becoming an underdeveloped nation in scientific research," pervaded reports and minutes of ministerial meetings.[35] Comparisons with the American budget devoted to scientific research were also common, as well as calls to articulate academic research with industrial production, following the American strategy designed to boost national economy through scientific and technological development. However, as historian Nicolas Simoncini has highlighted, this perception hardly corresponded to reality. Between 1954 and 1959, the French economy thrived: its GDP increased by 41 percent as exports grew by 44 percent and industrial production by 47 percent.[36] But even though the loud refrain of falling behind did not accurately reflect reality, it had a powerful hold in political circles as a performance of French independence.

With the start of the Fifth Republic, scientific and technical development were set to serve state interests via a new governmental structure to endorse so-called basic research. An Interministerial Committee for Scientific and Technical Research was tasked with coordinating research at the highest governmental level, defining the orientation of France's science policy at large scale. It provided advice to a specially designated executive body— the General Delegation for Scientific and Technical Research, or DGRST, which decided on the technoscientific fields likely to boost economic, political, military, or social goals. Their applied usefulness was important, as was their state of development relative to other countries.[37] The exploration of the oceans immediately stood out for its international, economic, and geopolitical relevance: Under the label "Exploitation of the Oceans," the political goal of investing in basic research was made explicit. According to delegates at the DGRST, marine sciences were significant both to "scientifically explore an unknown environment that covers two-thirds of the globe," and for the "direct benefits" that could be obtained in weather forecasting, navigation, enhancing fisheries, military defense (specifically in collaboration with NATO), exploring "new oceanic industries" (like the exploitation of bromine, magnesium, seaweeds, or hydrocarbons), and managing coastal pollution. All of these were activities that other nations were already pursuing in the oceans, and that the 1958 convention on the Law of the Sea regulated in the new legal framework.[38]

In line with the motive of lagging behind, the DGRST devoted a special budget aimed at—literally—fixing the "delay of France in these domains" through developing fundamental research and logistical, technological, and human resources. The case of high seas fisheries showcases the interministry plans to keep up with the Joneses while displaying a geopolitical agenda. By the late 1950s, a number of nations had succeeded at articulating the importance of basic research in their fishing industry, which had growing political and economic significance. Japan counted three universities devoted to fisheries oceanography, the Japanese pearl industry employed more than 500 researchers, and marine biologists were appointed to lead Japan's fishing fleets. In the Soviet Union, the Russian Federal Research Institute of Fisheries and Oceanography (VNIRO) gathered about 2,600 researchers, most of them experts in chemistry and marine biology; while the US had more than 2,000 experts devoted to applied oceanography (including fisheries). More modest in its resources, the UK possessed a laboratory specialized in fisheries that employed 180 researchers.[39] In France, a similar strategy could be read between the lines in the interministry wish to boost ocean research. Roger Pacque, secretary at the prime minister's cabinet, observed that France needed to invest in oceanography to strengthen its presence in distant seas, where it held political interests and influence. Those regions, euphemistically termed by Pacque as "particularly attached to the French culture," included former or current French colonial territories: the Indian Ocean surrounding Madagascar, the southwestern Pacific, and the Antilles, plus Brazil and Argentina.[40] In the framework of the new legislation on the high seas, which established freedom of fishing in international waters, scientific surveys in regions where French fishing fleets were already present (the Moroccan Atlantic, English Channel, North Sea, and the Antilles) could enhance France's control over international waters where its fleets deployed their nets.[41]

A TOOL TO BOOST FRENCH MARINE SCIENCES

Ministerial expectations were channeled through a newly created expert committee, the Committee "Exploitation of the Oceans" (COMEXO). Inaugurated on January 11, 1960, COMEXO gathered a group of prominent experts in marine sciences handpicked to decide on the orientation of the national program. At COMEXO's meetings, the three main actors presented so far sat together to discuss the future of France's ocean sciences: Jacques Bourcart, recently retired from his teaching appointments, represented the community of marine geologists. Jacques Cousteau participated as a prominent developer of underwater apparatus and pioneering research, and

Henri Lacombe voiced the stand of physical oceanographers. Besides them, the bias toward fisheries—the most important natural resource France was exploiting from the oceans—was clear in the group's composition: the director of the Technical Institute of Maritime Fisheries, Jean Furnestin, the vice president of the Office of Scientific and Technical Overseas Research, Paul Budker, and the president of the Confederation of Fisheries' Shipowners, Ferdinand Sarraz-Bournet, were present to voice the concerns of the industry, while three marine biologists backed them with their scientific expertise. Hydrographer and NATO-collaborator Marc-Marie Eyrès ensured a smooth connection with NATO military-scientific surveys, while explorer Theodor Monod provided external insights. The limited representation of marine geosciences on the committee can be attributed to the fact that the production of offshore hydrocarbons was not yet among the French government's priorities, as the next chapter will detail.

COMEXO was conceived as a temporary committee that, over four years (between 1961 and 1965), would identify the most acute needs in the marine sciences, establish an annual program of cooperative research among national laboratories, and monitor its proper development.[42] For those purposes, the DGRST devoted the equivalent of 8.65 million US dollars for the four-year period, the most well-funded program on its list.[43] Under the guidance of COMEXO scientific experts, the budget was invested to address the three weakest points of French marine sciences: providing research institutions with adequate technologies and equipment, organizing joint missions to prevent redundant work, and designing an official training plan in marine sciences for graduate researchers (table 1.1).[44]

Table 1.1. Budget allocation planned by COMEXO in 1960 for a four-year period (including investments and functioning costs)

Institution or disciplines	Budget to be invested (in 1960 US dollars)
Applied oceanography (fisheries)	126,305
Physical oceanography	760,000
Marine biology and geology in university labs	610,004
Marine biology[a]	408,356
International cooperation	857,446
Oceanographic vessels and submersibles	519,446

The laboratories in marine biology and geology are not specified.
[a] Including five marine stations and the National Museum of Natural History.
Source: DGRST, "Rapport du Comité d'Études 'Exploitation des océans,'" March 12, 1960.

At COMEXO's behest, marine geology (that is, the exploration of the seafloor) was identified as a branch of marine sciences that deserved special attention, next to physical oceanography and marine biology.[45] The move held no precedents, since until that time geology was a research field restricted to the mainland—with the notable exceptions of some enthusiastic pioneers, like Jacques Bourcart. As table 1.1 shows, it was not yet recognized as a field as important as physical oceanography or marine biology with industrial applications; but it was gathered under the same umbrella as marine biology at university laboratories.

COMEXO was designed as a scientific committee with decision-making power. The DGRST credited the group with this ability, acknowledging the pivotal role of basic research in France's marine agenda. As the DGRST's delegate Louis Jacquinot affirmed, the delegation firmly believed that the future was "determined by fundamental and basic research."[46] Behind this affirmation lay the rationale for the so-called linear model of innovation, in which basic science led to useful knowledge, which in turn enabled the development of new technologies and the enhancement of industrial processes.[47] On the other hand, the emphasis on promoting basic research was a central part of the rhetoric displayed by numerous national governments (notably the US), which saw in academic research a tool to ease diplomatic tensions through international cooperation and a means to approach and learn about the technoscientific capabilities of other nations.[48] The committee's actions, thus, aimed to promote fundamental studies at universities and public research laboratories that would constitute the pillars of future applications in navigation, exploitation of marine resources, or military surveillance.[49]

Despite the fact that COMEXO's decisions had to be framed within the guidelines defined by the Interministerial Committee and the DGRST, this group of marine researchers demonstrated their capacity to shape the political strategy on marine sciences. This was a notable power that, as we will see, would not continue once the seafloor became a site of major economic concern. Before moving to the seafloor's exploration and marine geosciences, it is instructive to learn about how COMEXO shaped the national strategy to explore the new, underwater territory by distributing its efforts between national research and international cooperation.

As detailed above, the French government was especially keen on strengthening France's presence in the high seas for fishing and geopolitical reasons. However, COMEXO experts deemed it unfeasible to start a national strategy by moving research resources far from home. They argued that France did not yet possess the human and material resources required to take advantage of long campaigns across the high seas. No matter how significant political interests were, to mobilize human and logistical resources to distant regions would be extremely costly in terms of time, money, and effort, while reported benefits would be scarce. The study of the distant seas could be pursued during international cooperative projects in which Lacombe and the other members of the committee would continue participating. Conversely, devoting the first years to exhaustively exploring the coastal waters of continental France would not only contribute to training experts in new research techniques and methodologies but would also result in detailed inventories of France's underwater territory, including its potential natural resources. As the COMEXO members argued, studies on the English Channel, the North Atlantic, and the western Mediterranean could reconcile national with scientific interests, did not require an excessive investment of money, and could be started with available resources.[50] In this sense, the exploration of the western Mediterranean's seafloor seemed an obvious choice for a starting point. COMEXO's budget and priorities translated into opportunities to introduce marine geophysical devices as methods for exploring the seafloor, allowing French geologists to delve into its vertical dimension for the first time.

THE MEDITERRANEAN HUB: MARINE GEOPHYSICS, FROM EWING TO COUSTEAU

From his seat at COMEXO, Jacques Bourcart considered the newly generous DGRST budget on ocean research as a unique opportunity to move the seafloor's exploration into an internationally competitive level: The time was ripe to push for acquiring marine geophysical technologies, through which the vertical dimension of the seabed—its layered strata, composition, and features—could be studied.

When COMEXO was inaugurated, the most capable oceanographic nations were rapidly incorporating marine geophysical surveys into the seafloor's geological study.[51] Marine geophysics involved the study of physical properties, like seismic waves and magnetic fields, to understand the seafloor's processes and deep structure.[52] In contrast, marine geological techniques like coring or dredging focused on examining sediments, fossil records, and topographic formations to interpret the seafloor's evolution. While in France geophysical methods were absent from marine geologists' toolbox, in the US, where these devices were first conceived, oceanographers had already launched ambitious cruises to understand the seafloor's composition around the world. The development of seismic techniques for marine geophysical research provides a glimpse of the divergent pathway that the exploration of the sub-seafloor was taking on the other side of the Atlantic Ocean.

Originally conceived at the turn of the century by private American companies for shallow-water oil and gas exploration, marine seismic techniques were first adopted for scientific research in the 1930s.[53] Maurice Ewing, a physicist at Lamont in the US, pioneered the use of seismic reflection to study the deep seafloor's structure (fig. 5).

Ewing was introduced to marine geophysics while studying physics at Rice University. During the summers, he worked with an oil prospecting crew in the shallow lakes of Louisiana, where they conducted seismic and gravimetric surveys to identify buried salt domes—key indicators of hydrocarbon deposits—along the Gulf Coast. Ewing's growing interest in this field led him to write a PhD thesis on the theoretical physics behind seismic reflection techniques.[54] In 1935, established as assistant professor at Lehigh University, Ewing began applying the principles of shallow-water oil

Figure 5. Marine geophysicist Maurice Ewing, pictured at the Lamont Geological Observatory. The long paper sheets spread across the desk are seismic profiles, visual representations of seafloor structures and composition created using seismic reflection or refraction data. Date unknown. Source: Photo by Warman, Columbia University, gift of Ron Doel. Niels Bohr Library and Archives, American Institute of Physics. Reprinted with permission.

prospecting to the exploration of the oceanic crust. This technique involved generating sound waves near the surface, which then traveled through the seafloor's deep layers. A series of towed hydrophones on the surface recorded the reflected waves. Because waves propagate at different speeds through different sediment types, the hydrophones registered varying arrival times. By analyzing these differences and knowing the propagation velocities, he could infer the sediment types that made up the seafloor and estimate its depth. With funding from the Geological Society of America and logistical support from the US Coast and Geodetic Survey, Ewing demonstrated that seismic techniques could be applied to identify formations and specific geological horizons up to two hundred meters deep. Accompanied by a team of young physicists, Ewing progressively moved his surveys from the continental shelf to deeper waters, while testing the use of TNT explosives and other methods to create more powerful sound signals. His innovative results led him to win the support of the US Navy during World War II and, afterward, to secure generous funding from the Office of Naval Research. The use of marine geophysical devices in sciences jumped to the other side of the Atlantic Ocean in 1937, when Ewing began to collaborate with British geophysicist Edward Bullard, a researcher at Cambridge University. Other American research teams followed Ewing: at the Scripps Institution of Oceanography, physicist Russell W. Raitt began in 1947 to study how sound waves are reflected from within the seafloor, using large charges of dynamite to create acoustic signals (figs. 6 and 7).[55]

In these early years, marine geophysics was frequently separated from marine geology in terms of researchers, expertise, and expeditions, but that did not mean collaborations were nonexistent. Francis P. Shepard and Raitt, for instance, moved to the Scripps Institution of Oceanography almost simultaneously, and it is likely that they shared their research outcomes about the ocean floor with each other.

In the US, collaborations between geologists and geophysicists began to take a more institutionalized shape during the 1950s, when seismic devices were reliable enough to produce seismic profiles that needed to be interpreted. Dynamite—dangerous to carry onboard, extremely harmful for the marine environment, and expensive without the US Navy's support—was exchanged for innovative and more affordable means.[56] John B. Hersey, at the Woods Hole Oceanographic Institution, tested a sounding source in 1952 that generated acoustic waves from a high-voltage electric spark discharged underwater, a type of seismic source known as a *sparker*.[57] Simultaneously, Maurice Ewing became the first director of the newly founded Lamont-Doherty Observatory, where he developed a sounding source that generated sound pulses by discharging high-pressured air, the *air gun*.[58]

Figure 6. American physicist Russell W. Raitt throwing a TNT charge during the Capricorn Expedition, 1952. This method was widely used in academic marine geophysical expeditions to generate powerful acoustic waves before the development of more sophisticated, and less aggressive, devices. Source: Special Collections and Archives, UC San Diego, La Jolla. Reprinted with permission of UCSD.

Figure 7. TNT seismic explosion during the Woods Hole Oceanographic Institution's Atlantis Cruise 151 in the Mediterranean, 1948. The near-surface explosions created a violent shockwave, producing a visible splash and a temporary cavity in the water. The technique was widely used among marine geoscientists in the early days of marine seismic research. Photo by Don Fay, © Woods Hole Oceanographic Institution. Reprinted with permission of WHOI.

Equipped with enhanced geophysical techniques and generous economic support from the US Navy, Ewing and his team began to undertake large geophysical surveys around the world, which led to unprecedented discoveries about the ocean floor. In 1955 Ewing demonstrated that the oceanic crust was thinner that the continental crust, and that the thickness of accumulated sediments was also thinner in the deep oceans.[59] This evidence suggested that the ocean floor was much younger than the continents, thus offering hints that the key to understanding Earth's dynamics was to be found at the bottom of the oceans. This hypothesis was reinforced with parallel scientific discoveries backed by military data and sounding devices. At Lamont, geologist and research assistant Marie Tharp had been analyzing echo-sounding profiles from around the globe, finding that underwater ridges similar to the Mid-Atlantic Rift (discovered during the *Challenger* expedition in 1872–76) were ubiquitous, whereas her collaborator Bruce C. Heezen identified that the epicenters of earthquakes coincided with the ridges.[60] Suspecting that the mid-ocean ridges could correspond to extensional regions (areas where the ocean crust is pulled apart), Ewing undertook in 1959 an around-the-world cruise to study ocean ridges with geophysical techniques.[61] These and other geophysical and magnetic evidence pointed to the existence of continental drift, giving impetus to American oceanographers to mobilize their vessels and seismic techniques to explore the seafloor around the globe.

As we have already seen more than once, American oceanographic capabilities sharply contrasted with the resources available in France. While Ewing was leading months-long surveys across the world oceans, probing the seabed's deep composition, in the French Mediterranean, Bourcart was organizing daylong cruises to survey the seabed's topography up to the continental shelf's limit. Only those French researchers connected to American partners and supported by private funding had been able to start experimenting with geophysical devices at the time the French government launched major support for ocean sciences. The unique case of Jacques Cousteau and his team, at the Museum of Monaco, was central to introducing France's community of marine geologists to geophysical techniques for exploring the western Mediterranean.

Jacques Cousteau's relationship with American marine geophysics was born from a fortuitous encounter: In 1952, the National Geographic Society introduced to him MIT engineer Harold Edgerton. The meeting happened after the *Calypso*'s inaugural expedition to the Red Sea (when Cousteau famously said, "We must go and see"). The mission had pursued the dual goal of photographing red corals while proving to oil companies that the *Calypso* and its divers could explore the seafloor for petroleum and mineral deposits. In those early years of seafloor exploration, collaborating with oil firms was regarded as an ideal strategy for funding their filming cruises around the world. By traveling to New York, carrying under his arm a folder containing the pictures acquired in the Red Sea, Cousteau was seeking one more patron. He planned to request funding from the National Geographic Society to continue producing photographs and film footage that could show the marvels of the deep ocean to a wide audience. Interested in Cousteau's proposal, the society agreed to provide the Monaco team support to overcome the main obstacle of underwater photography, the scarcity of light, by introducing Edgerton to Cousteau. This marked the start of a life-long friendship (fig. 8). Edgerton, known for his studies on stroboscopic lighting (which famously allowed him to photograph the impact of a drop of milk), produced underwater flashes and cameras for Cousteau's group, earning the nickname "Papa Flash" from the *Calypso*'s crew (his son Robert, who helped him as an assistant, became "Petit Flash," or *Little* Flash, in English).

Edgerton's underwater photo devices relied heavily on using them in conjunction with technologies designed for seafloor exploration. To precisely detect where in the murky depths on the seabed cameras and ancillary instruments were to be located, Edgerton developed the "pinger" from existing sonar systems. As Cousteau's interest in filming underwater wrecks intensified, Edgerton produced the *boomer*, a sounding source whose waves penetrated below the first meters of the seabed's sediments. This instrument became as useful for searching for the ancient Greek city of Helice as for conducting geological research.

The *boomer* became a primary tool for conducting marine geophysical surveys across the Mediterranean Sea onboard the *Calypso*, but also for other American research teams with whom Edgerton collaborated (fig. 9). American interest in a region so far from the US coastline resides in that the Mediterranean basin's northern shore constitutes an excellent setting to test and explore the tectonic movements that generate folding mountain chains, like the Alps and the Apennines.[62] Oceanographer John B. Hersey, from the American Woods Hole Oceanographic Institution, for example, started collaborating with Edgerton and his *boomer* in 1958 to study the processes that transform Earth's crust in the Mediterranean basin. Through Edgerton, Cousteau, and Hersey, a strong connection was forged between American marine geoscientists and the community of ocean experts in Monaco. The Mediterranean spot became a meeting point for scientists interested in innovative oceanographic technologies, offering new opportunities for young experts in France and Monaco to learn from the American techniques and to set sail with them.[63]

Figure 8. Captain Jacques Cousteau (left) with Harold Edgerton, photographed onboard the *Calypso*. The two men collaborated closely in the 1950s, with Edgerton's high-speed photography and strobe light technology enhancing Cousteau's underwater explorations and bringing greater visibility to marine life and oceanographic research. Source: MIT Museum, Edgerton Digital Collections. Reprinted with permission of MIT Museum.

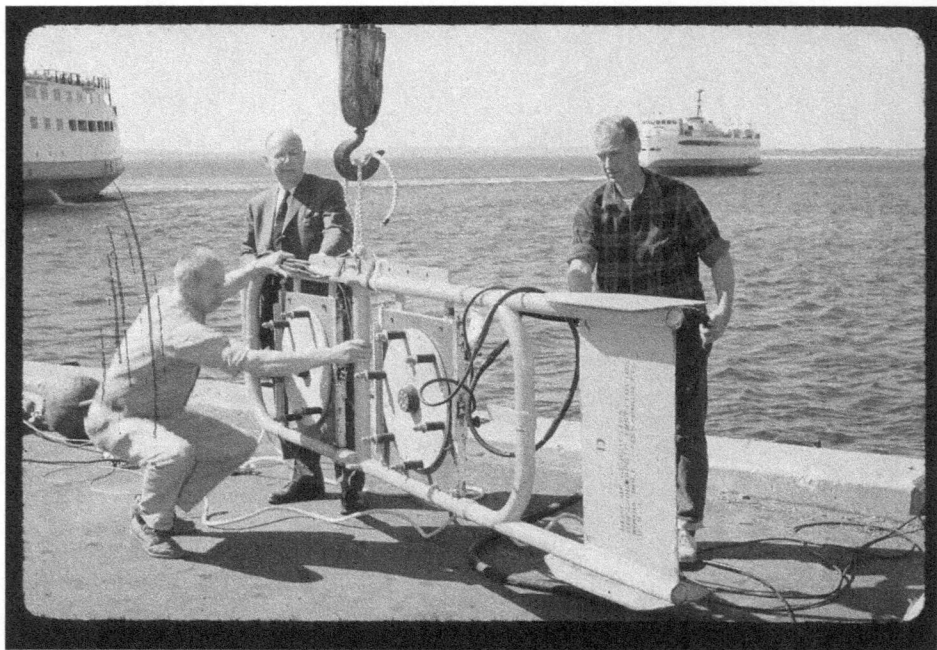

Figure 9. Harold E. Edgerton (center), Bill MacRoberts (right), and an unidentified man pose by a large sonar boomer, on a dock at Woods Hole Oceanographic Institution in 1964. The boomer was a marine geophysical tool that, instead of TNT explosions, generated a burst of energy to create a sounding wave used to produce high-resolution images of the seafloor. Source: MIT Museum, Edgerton Digital Collections. Reprinted with permission of MIT Museum.

The numerous scientific activities conducted onboard the *Calypso* demonstrate Captain Cousteau's strong interest in (or, at least, important contribution to) marine geosciences. This aspect has received limited attention in his biographical narratives, which usually focus on his trajectory from naval officer and explorer to famous underwater filmmaker and conservation advocate.[64] While Cousteau played a significant role in bringing the wonders of the oceans to wider audiences, he was equally important in channeling the marvels of the Mediterranean seafloor to the French scientific community. However, the materials he presented to this later audience were different from the lively images that captivated the broader public. During scientific conferences and workshops in France, Cousteau presented sounding data and seismic profiles acquired onboard the *Calypso*, which together displayed the seabed's topography and stratigraphic composition. There are also signs of Cousteau's close relationship with Jacques Bourcart. As mentioned earlier, Bourcart and Cousteau first encountered

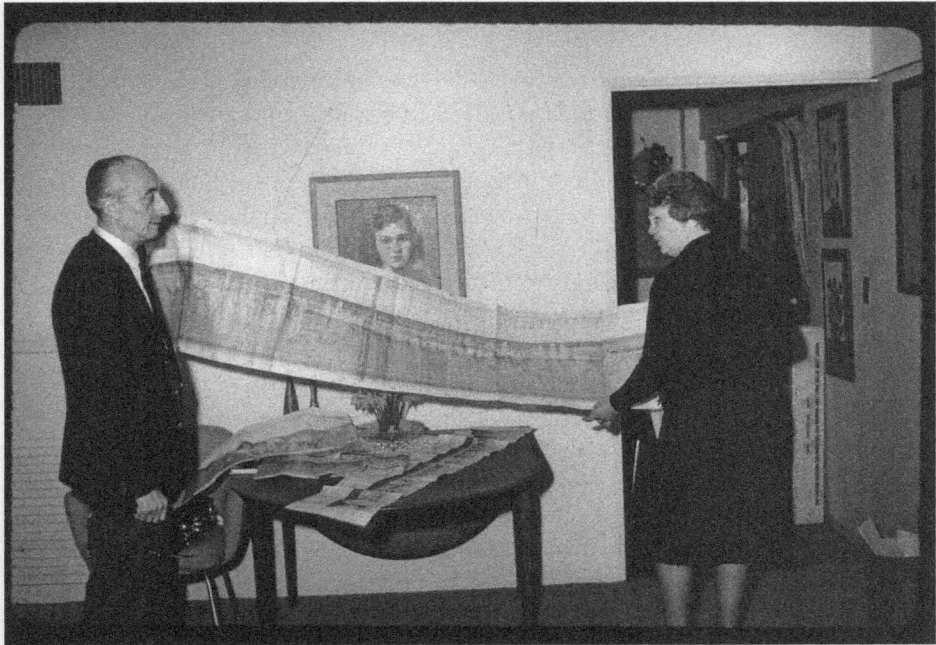

Figure 10. Jacques Cousteau (left) and Esther Edgerton (right) holding a seismic profile produced with a sonar sparker in the Mediterranean Sea; likely at H. E. Edgerton's home in Cambridge, Massachusetts. The seismic profile, depicting the Ligurian seafloor, shows some elongated structures that correspond to salt domes. Source: MIT Museum, Edgerton Digital Collections. Reprinted with permission of MIT Museum.

each other at the French Navy's expert Committee of Oceanography and Coastal Studies in 1946. At the same time, they became onboard collaborators at the *Aviso Ingénieur Elie-Monier*, the ship Cousteau commanded at the Hydrographic Service, from which Bourcart conducted his first marine geology studies.

Cousteau's scientific results became the spark that ignited the flame for organizing a national-scale marine geophysical survey of the western Mediterranean with COMEXO's funding. In 1961, Bourcart gathered at Villefranche-sur-Mer the community of French geologists, physicists, hydrographers, and engineers interested in exploring the seafloor. Among talks on coastal and underwater sediments, Cousteau showed the audience the latest sounding profiles acquired onboard the *Calypso* across the Ligurian Sea, facing Monaco, by using Edgerton's echo-sounders.[65] In this region, his team had encountered small protruding mounts, now labeled simply "landforms," whose origin was completely uncertain (fig. 10).[66] After discussing these results with Hersey, Cousteau announced a joint geophysical survey of the region to study the sub-seabed structure of those formations and investigate their origin.

The possibility of conducting parallel investigations enthused Bourcart and the attendees. From their perspective, acquiring marine geophysical systems could align smoothly with COMEXO's (and the government's) priorities. Marine geophysical expeditions would produce novel information about the Mediterranean basin while directly addressing the government's mandates: to broaden understanding of the potential natural resources beneath the national submerged territory and to avoid falling behind internationally in oceanographic capabilities. Under Bourcart's chairmanship, attendees agreed on investing COMEXO's budget in a cooperative program to survey the western Mediterranean seafloor with geophysical tools while pooling vessels, experts, and laboratories.

When Bourcart informed the DGRST of the meeting's decisions, the reasons offered to promote the program focused on geopolitical interests rather than the scientific ones. In a report transmitted to the DGRST delegates, Bourcart revived the motive of promoting sciences to increase national prestige and fanned the fear of lagging behind other countries in seafloor exploration. He alleged that investing in a cooperative, large national program of marine geophysical surveys was essential to keep pace with American, German, and Soviet research teams, which were already surveying the western Mediterranean with their large oceanographic vessels and well-prepared research teams—alluding to the surveys that Hersey and other foreign research teams were conducting.[67]

The introduction of marine geophysical techniques to the community of French geologists offers evidence of the unprecedented freedom COMEXO possessed to shape the political agenda on marine sciences, precisely because of the flexibility the DGRST had granted it to manage its budget. Marine geophysics was not among the DGRST's guidelines; it was not required for enhancing high-seas fisheries, and the French Navy was already developing its own tailor-made technical expertise. However, COMEXO experts were aware that developing academic marine geophysics in France was essential to stay competitive internationally, especially if the country aimed to conduct pioneering investigations of the western Mediterranean.

With COMEXO's economic support, French marine geophysical surveys were launched in January 1962. Experts from the University of Marseille and the Musée de Monaco led a pilot survey utilizing refraction seismic techniques across the Gulf of Lion, onboard Monaco's vessels *Espadon* and *Winaretta Singer*. By detonating five hundred kilograms of dynamite in thirty-six shots, they produced the first geophysical profiles of the seafloor's layered structure (although with coarse resolution). These were coupled with geological samples recovered by the team of geologists at Villefranche-sur-Mer.[68] In parallel, COMEXO granted Cousteau's team a budget of 165,000 US dollars to undertake a three-year program to survey and chart the western Mediterranean with reflection seismic techniques.[69] Through these missions, the seafloor began to take shape as geologists glimpsed its sedimentological horizons, from the solid granite basement to the upper soft and unconsolidated sediments.[70] Yet the results were not sufficient to offer definitive evidence of the Mediterranean's deep composition. Because of their head start and large-scale campaigns, American researchers were the first to point out the Mediterranean's hidden treasures.

THE MEDITERRANEAN'S NEW DIMENSION
AND THE GAP BENEATH

In an unexpected turn of events, the protruding mounts Cousteau had casually glimpsed over the seabed turned out to be major hints of hydrocarbon deposits. In April 1965, during the Seventeenth Symposium of the Colston Research Society at Bristol, Hersey summarized the last decade's results he and his team had acquired studying the Mediterranean, including cooperative missions with Monaco's team.[71] After characterizing the structure of its deep seafloor, detailing its stratigraphic sequence, drawing correlations between the western and eastern basins, and mapping its morphology, Hersey mentioned that, in geophysical profiles acquired with reflection seismics in the Balearic basin, his team had identified "dome-like structures" that correlated with Cousteau's protruding mounts. Most likely, he added, those were salt domes, because evaporites (salt minerals) are commonly found across Italy's coastline and they could stretch out beneath the seabed.[72]

Hersey's discovery had a major impact on both the European geological community and on the oil industry for the next decade. Salt layers and domes are structures constituted by salts that accumulate after massive evaporation of saltwater, as happens in natural saline flats. At the scale of geological history, salt (evaporites) accumulations tend to originate when ancient oceans open or disappear. Salty minerals, which accumulate at the basins' bottom in horizontal layers, are covered during the following millennia by more recent sediments. But since salt is a very plastic material, less dense than the sediments covering it, it tends to ascend to the surface, deforming overlaying layers. This process creates salt domes, elongated structures made out of evaporitic materials. In addition to their geological interest, domes and layers of evaporites were also in the spotlight for oil companies. Due to their high impermeability, evaporites constitute effective seals for hydrocarbons, which tend to accumulate below. This relationship had first been proven on dry land in the late nineteenth-century Louisiana oil fields, and from that time salt domes constituted benchmarks for oil exploration, on emerged lands as well as offshore.

Henry W. Menard, from the Scripps Institution of Oceanography, had simultaneously found equivalent evidence in the late 1950s while studying other research topics using different techniques—sedimentation and underwater currents around the Rhône's deep-sea fan, in the Balearic basin's abyssal plain.[73] While sounding the sedimentary fan's morphology with echo-sounders, Menard and his team had encountered a dozen knolls protruding between twenty and more than a hundred meters up from the seabed. That study revealed that the formations looked exactly like the small mounts found at Sigsbee Knolls (in the Gulf of Mexico) that American geophysicist Maurice Ewing had identified as underwater salt domes.[74]

The potential existence of salt domes in the western Mediterranean sparked the immediate interest of French geoscientists, since they could support Bourcart's hypothesis of a major seawater regression. By studying the timing when the thick evaporitic layer had deposited, through which processes, and how the Mediterranean basin looked at the time, geologists could understand the region's evolution. If salt domes were as ancient as the basin, it would imply that evaporites had accumulated from the closing of the former, wider Tethys Ocean; but if they were more recent, it could indicate massive events of seawater regression in very recent times, when the Mediterranean already existed in the shape we now know. Yet geologists were not the only ones striving to define the age and composition of those formations. As we will see, French oil companies rushed to acquire exploration leases across the western Mediterranean, driven by the possibility that those were indeed salt domes.

Menard and Hersey's investigations across the Mediterranean evidenced, first, how involved American researchers were in the region; and, second, how large the research gap was between American and French geoscientists. American geoscientists presented results acquired during long offshore cruises over many years, combining geological and geophysical techniques with cutting-edge technologies. In contrast and despite the recent French enthusiasm for marine geophysics, COMEXO had just began to coordinate efforts between geologists and geophysicists, surveys had only been conducted across the shallow continental shelf, and French researchers did not even yet have access to an operational oceanographic vessel from which to conduct large-scale expeditions.

FRANCE'S JOURNEY INTO SEAFLOOR
RESEARCH

Nations don't follow a single pathway when committing to a research field. The motivations driving these decisions shape not only the direction the research takes but also reveal something deeper about the state: its broader priorities, aspirations, and self-perceived place in the international landscape. The first steps of government-supported seafloor exploration tell us as much about the development of marine geology as they do about France itself. They reflect a dynamic interplay of national ambition, international developments, and a blend of new and old scientific interests in the oceans.

In the late 1950s, marine geology in France was not directly linked to national energy interests. Instead, it was shaped by broader political and scientific aspirations. De Gaulle's ambition for the Fifth Republic extended to the seas, as ocean exploration became part of a larger vision to assert France's position in the global order. The declinist discourse of lagging behind foreign countries fueled a desire to boost national prestige and attain geopolitical goals through targeted investments in science. At the same time, shifts in the international attention to and conflicting regulation of the oceans heightened the importance of understanding the seafloor. This exploration, however, required far more than geological expertise. It demanded advanced research technologies, like marine geophysical systems, increased funding, and collaboration across scientific disciplines. Under COMEXO, geologists joined forces with other marine scientists, sharing resources—like oceanographic vessels—and research spaces.

The story of COMEXO and its investment in marine geophysics captures a period when academic researchers had greater autonomy in shaping the national research agenda, compared to the years that immediately followed— when the seafloor became France's new energy frontier. At COMEXO's time, marine geosciences were seen as part of a broader scientific domain, not focused on specific applications like hydrocarbon or mineral exploration. This had its drawbacks: marine geology was not as high a priority for the government as other marine sciences, such as physical oceanography and marine biology, which attracted attention for their military or commercial relevance. However, this broad perspective on the usefulness of marine geology allowed researchers to design a general national plan for seafloor research and to allocate the budget without the pressure of immediate economic goals.

For the seafloor's exploration, the declinist discourse of lagging behind gained particular traction among France's political elite only when the national oil industry turned to the seafloor as a critical area for securing future energy supplies. Only then, when its potential value became undeniable, did seafloor exploration become a major political concern.

[CHAPTER 3]

France's New Economic Frontier

There is something in a treasure that fastens upon a man's mind.
JOSEPH CONRAD, *NOSTROMO*

In February 1964, just as the expert group COMEXO was reaching the end of its first term, the DGRST justified its continuation, pivoting on the assertion that the oceans were "no longer mere surfaces of communication." Now, from its perspective, the oceans had become environments where the "scientific adventure precedes a rational exploitation of the wealth contained in the mass of waters or in the seabed."[1] In barely five years, imagined underwater treasures had become a fixed picture in the mind of French politicians, diplomats, and industry experts. Visions of treasure added to the hitherto prevailing rationale of exercising geopolitical control and enhancing France's global reputation through ocean sciences. Events happening far away from the oceans prompted this change in priorities—a transformation brewing in the former colonial territories of the North African desert.

The perception of the marine environment shifted due to the crumbling of France's colonial empire and the resulting instability of hydrocarbon suppliers. France's oil industry turned to the seafloor as "the only virgin territory where France's oil industry could develop," an outlook that had major consequences for the political organization of ocean research.[2] Oil industry experts and civil servants laid the groundwork for the future of ocean sciences in an attempt to connect academic research to industrial activities. The foundation of the Centre National pour l'Exploitation des Océans (National Center for the Exploitation of the Oceans, or CNEXO) marks a turning point in the history of France's ocean exploration. It was established in 1967 as a management office at the elegant Avenue de Iéna in Paris, surrounded by private residences of ambassadors, politicians, and diplomats. It would not establish as a major scientific hub until 1971, when CNEXO's Oceanological Center was built on the Atlantic coastline at Brest. Regardless of its initial administrative role, CNEXO's creation was laden with symbolism for France's ocean sciences. Its institution represented the culmination of a process through which the hydrocarbon industry associated with scientific research and, at the same time, set a departure point for France's stance as an international oceanographic power. This chapter traces the pathways that converged toward CNEXO's creation, from the processes to set up a national offshore hydrocarbon industry to the explicit emphasis on seafloor exploration in France's ocean policies.

THE OIL INDUSTRY'S "PLAYGROUND," FROM THE DESERTS TO THE SEABED

In 1958, as Charles de Gaulle took office and UNCLOS I took shape, the French oil industry faced a recurrent problem—the need to relocate its efforts in order to secure France's energy supplies. At the French Institute of Petroleum (IFP), Joint Director André Giraud witnessed with dismay the collapse of the national oil industry in North Africa: France's strategies to achieve energy independence by relying on these regions suddenly vanished.

Giraud, a thirty-three-year-old civil servant, was committed to helping the nation acquire its long-awaited energy self-sufficiency, a goal that aligned with the new government orientation.[3] First trained as an engineer of the Corps des Mines[4] at France's prestigious École Polytechnique and École Nationale Supérieure des Mines, Giraud specialized as a petrochemical industry expert in Texas. His career took off in 1951 at the IFP, a public institution devoted to offering technical, scientific, and economic assistance to French oil companies, where he swiftly ascended from researcher to the position of joint director in just seven years.[5] In that period, he contributed to the national efforts to overcome France's heavy energy dependence on foreign oil suppliers.[6] Yet, at the end of the fifties, the challenge appeared more daunting than ever.

To understand Giraud's concerns about the industry in the late fifties, we must go back to the end of World War II. After that conflict, France had secured its supplies in the Near and Middle East, which by then represented 90 percent of its imports.[7] But its suppliers proved unstable during the Suez Crisis of July 1956, when Egyptian President Gamal Abdel Nasser nationalized the British and French-owned Suez Canal Company. The subsequent Arab-Israeli war—with the involvement of France and the UK—prompted Nasser to close the Suez Canal from October 1956 to March 1957, affecting global commerce and interrupting the flow of oil and gas from the Middle East to European countries. New discoveries in French Algeria sparked hopes to secure a supply of hydrocarbons closer to the metropole. In 1956, the national French Petroleum Company (CFP) identified the oil reservoir of Hassi Messaoud and the gas field of Hassi R'Mel. These discoveries spawned a new political strategy, in which France would base its energy policy on the *pétrole franc*—the Algerian reserves.[8] Yet, at the turn of the decade, the process of decolonization incited experts at the Fuels Directorate, a division of the Ministry of Industry, to rethink this strategy. In July 1958, a coup d'état in Iraq led to the nationalization of its Iraq National Oil Company, which then expropriated foreign oil firms—including the CFP—of 90 percent of their exploration and exploitation leases. By 1960, fourteen sub-Saharan territories had gained independence, including oil-producing countries such as Nigeria and Congo-Brazzaville. Initially, the CFP went on with normal operations in these regions, as local institutions required training and expertise from French engineers. However, managers at the CFP realized that it was a matter of time before these regions followed Iraq in nationalizing their oil industries.[9]

The creation of the Organization of the Petroleum Exporting Countries (OPEC) in 1960 reinforced these anxieties. Experts at the Ministry of Industry feared that the organization would become a cartel in control of the oil market at a moment when imports from Algeria were not enough to supply the hexagon. In 1957, France had imported a mere 1 percent of the hydrocarbons it consumed from North Africa, but by 1960 this share had increased to 15 percent.[10] When Algeria obtained its independence in 1962, future prospects were dismal. Although the Evian Agreements ensured that the CFP could still operate in the territory under the same conditions, disagreements between the Algerian and French governments over oil exploration, production, refining, and transportation multiplied.[11] Fearing to jeopardize its future energy supplies, the Ministry of Industry and French oil companies decided, in the words of the high official Gérard Piketty, to "seek a playground" for the national companies to ensure a secure supply for France and "nondependence" on the major cartel.[12]

The oceans then began to emerge as such a "playground," as Piketty had called them. An early orientation toward the seas seemed promising: the new legislation on the continental shelf ensured sovereignty rights to national governments for the exploration and exploitation of hydrocarbons as far as their technological capabilities allowed. In the US, offshore oil production across its continental shelf had increased from 133 barrels a day in 1954 to 444 in 1960, and constituted 6 percent of the country's oil production.[13] In Europe, offshore oil exploration took off in 1959, when Shell and Standard Oil found the biggest natural gas reservoir ever discovered in Europe off the Netherlands' coast.[14] In the Law of the Sea's framework, the North Sea's continental shelf was divided among its seven bordering countries, and the allocation of exploration leases was regulated. This facilitated the acquisition of leases by foreign oil companies for offshore exploration.

Against this backdrop and before coordinating a national plan, French public organizations and oil companies started to seriously consider potential developments at sea, starting to probe their capabilities in the shallow waters of the North Sea. France's Office of Petroleum Research, a public institution created in 1945 to coordinate and organize oil exploration in France's colonial territories, and its Autonomous Board of Petroleum, a national oil company created in 1939, undertook their first seismic surveys across the German, Norwegian, Danish, and Dutch continental shelves; and the CFP joined them one year later to explore the British underwater region.[15] Growing international interest in offshore oil provided a glimpse of the new geopolitical theater that, in the years to come, the submerged territory would become. Each nation's technological capabilities would decide the level of control it exerted over new sources of hydrocarbons and minerals.

The industrial effervescence off the North Sea at the turn of the sixties inspired public officers back in France. At the IFP headquarters in suburban Rueil-Malmaison, neighboring Paris, André Giraud enthusiastically took up the challenge of defining, as soon as possible, a national strategy in order to develop an efficient offshore oil industry. Giraud advocated for designing an "exploitation of the oceans" policy as a means to secure France's energy supplies. From his point of view, the oceans were "virgin" territories equivalent to the former colonial regions, and thus appropriate spaces for conquest, control, and exploitation. As he argued in a persuasive document to the prime minister's cabinet, postcolonial France would develop as a global power by expanding and exploiting new (underwater) territories—akin, as he specified, to the foundation of the greatest historical empires, the American and Soviet supremacy, or early modern colonial France. In this process, science and technology were called to play a crucial role. On land, zoologists, explorers, botanists, geologists, and engineers had built the foundations of colonial empires. Now marine scientists and engineers were essential to chart the underwater territory, identify natural resources, exercise territorial control, and develop means for industrial production.[16]

Giraud's discourse provides a first hint of what some historians and firsthand actors have called an oceanic neocolonialism: a world order in which power relations prevalent in colonial times would be replicated in the oceans.[17] Actors like Giraud believed that the oceans could replace colonial territories to supply the hexagon with minerals and hydrocarbons after a process of exploration, occupation, and conquest—a "conquest" that could only be achieved through technological innovation, essential to reach those deep, unknown, and invisible regions. Thus, from Giraud's perspective, the three pillars upon which the future national offshore oil industry would rely were technological development, the training of new experts, and their coordination under a state administration intent to reap the result of their efforts.

Other experts of the energy sector shared Giraud's convictions, like civil servant Maurice André Leblond, head of the Fuels Directorate, and Jean Blancard, the civil servant presiding over the Office of Petroleum Research. Giraud, Leblond, and Blancard envisioned the future of the French oil industry as being built upon the ocean floor.[18] In 1963, the three men agreed to implement, as soon as possible, a powerful and internationally competitive organization designed to advance a national strategy directing the oil industry's interests efficiently toward the oceans. In front of the French Council of Ministers, they pushed for the creation of a Committee of Petroleum and Marine Studies, or CEPM.

On March 26, 1963, the Ministry of Industry and the Ministry of Economic Affairs instituted CEPM, which gathered representatives from the main national oil institutions: the CFP, the Régie Autonome des Pétroles, and the IFP.[19] Giraud, recently appointed head of the Fuels Directorate, would lead the group.[20] CEPM was designed to create favorable conditions through which a national offshore oil industry could flourish by defining high-priority actions and organizing their execution, coordinating exploration and exploitation activities, promoting technological and logistic innovation, and stimulating collaborations between national institutions, thus preventing the unnecessary replication of studies. These actions would reduce the time and money invested.[21]

CEPM's uniqueness, in contrast to aggregations of oil firms in other countries, lay in the fact that its shape was closer to a national network than to a consortium of oil firms. Inside the oil industry, the creation of consortia was a classic strategy to explore the oil potential of new regions: a number of oil companies associated to share the economic burden of exploration and exploitation, while they also shared the outcomes of their efforts. Consortia were particularly suited for exploring offshore, as searching for oil was three times more costly at sea than across mainland territories, and the chances of finding a reservoir that could produce a return on investment were completely unknown.[22] Not many companies were prepared to risk their capital in developing exploration technologies, drilling new wells, or in undertaking exhaustive surveys without sharing the research burden with other firms. In the US, for instance, the main private oil firms devoted to offshore exploration (Continental, Union, Shell, and Superior Oil) had joined forces in 1946 under the CUSS Group to explore for oil off the California coast. Conversely, in France, CEPM was a structure tightly linked to the Ministry of Industry due to the public nature of its member institutions: the French government owned a third of the CFP, while the Autonomous Board of Petroleum was an entirely national firm. Both the Office of Petroleum Research and the IFP were public institutions directly dependent on the Fuels Directorate: whereas the first was devoted to drafting the national energy strategy, the IFP trained engineers and technicians and developed knowledge and new techniques to supply the national oil industry. CEPM's model, thus, ensured a smooth articulation between political priorities and oil exploration activities, avoiding frictions derived from the confrontation of private and public interests.

ENGINEERING AN UNDERWATER TOOLKIT
FOR THE NEW INDUSTRY

The first obstacle oil managers confronted at CEPM was the absolute lack of technologies and infrastructures for marine activities. Petroleum engineers had to learn how to conduct research at sea from scratch, first resorting to the navy's Hydrographic Service to request relevant information on meteorology, corrosion of materials, radiolocation systems, and research methods on gravimetry, together with bathymetric charts. Engineers were even invited to embark onboard the navy's vessel *Amiral Mouchez* to learn about techniques in gravimetry research.[23] However CEPM managers soon revealed their scant interest in the navy's gravimetry methods, because to spot potential hydrocarbon deposits the industry needed data with much higher resolution and detail.[24] Priority went to acquiring marine geophysical devices to enable researchers to *read* the seafloor's layered composition. Coring devices would allow the recovery of undisturbed samples from the seafloor, and research vessels would enable the oil industry to conduct its surveys without needing to subsidize external institutions.

Although foreign oil companies were already manufacturing their own marine geophysical devices, CEPM decided to design its own. In the short term of course this required a higher investment of time and money than acquiring foreign technologies, but in the long term it would free the French oil industry from reliance on foreign companies. This strategy aligned with the political ideology prevalent in Charles de Gaulle's administration, in which the overarching goal was to achieve France's energy autonomy and independence. The task of designing, building, and testing those new devices fell to the IFP, which inaugurated the Marine Project for those purposes. In 1963, large sums of money began to flow from the Ministry of Industry to the project: 9 million US dollars in 1963, 10.3 million in 1964, and 13.3 million in 1965.[25] Oil companies under CEPM also contributed financially to supporting projects of greater interest, which granted them the capacity to shape research according to their priorities. By contrast, in the academic sector, COMEXO, which devoted its budget to promoting marine research in universities, received a budget of only 8.65 million US dollars for a four-year period, to be shared by all marine disciplines.[26]

Engineers Jean-Pierre Fail and Jacques Cholet undertook the task of designing a high-penetration seismic device (that is, a geophysical technology capable of acquiring data from hundreds of meters below the seabed) that did not depend on exploiting large dynamite charges. Until the mid-1960s, large explosions of about ten kilograms of TNT per shot were an essential condition to obtain seismic profiles from the deepest layers of the seafloor. This method was the only known way to overcome the "bubble pulse": an air bubble generated when operating the sounding source that distorted acoustic signals.[27] However, seismic surveys with explosives were problematic: aside from the challenges of relying on large supplies of dynamite (not always available, and dangerous to have onboard), underwater explosions had sparked anger among fishermen along the Mediterranean coastline.[28] Each underwater detonation massacred all kinds of fauna, and a single survey involved dozens of explosions. Official documents from the oil company Entreprise de Recherches et d'Activités Pétrolières (ERAP) reported complaints of fishermen in the Gulf of Lion after they had run a seismic survey, while the newspaper *Le Monde* also reflected this concern in an article devoted to marine exploration.[29] Another technical challenge was the need for continuous detonations every few seconds to capture seismic profiles across the seafloor, rather than from isolated spots.[30] Without a continuous sounding source, the vessel couldn't move while collecting data, which resulted in both longer times at sea and higher costs.

Between 1963 and 1965, CEPM invested more than 300,000 US dollars in the development of seismic reflection devices at the IFP.[31] Fail and Cholet acquired a small tuna boat, the *Petite Marie Françoise*, and equipped it with radar, sonar, and a radio. From there, their team would be able to conduct as many tests as it wished without having to rely on external institutions and companies.[32] Next, they tested different systems of sound-wave generation at sea, like electromagnetic vibrations and compressed-gas explosions. After several trials, Fail and Cholet realized that, to overcome the bubble pulse problem, they could use small explosive charges detonated in a structure that would amplify the signal. Based on this premise, Fail and Cholet designed a brand-new device, patented in 1965: the *Flexotir*.[33]

The innovation was enthusiastically announced in *Le Monde*, which emphasized its environmentally friendly approach to offshore oil exploration with the headline, "The Flexotir, or How to Seek for Oil Without Killing Fish."[34] Unlike other seismic sources that relied on explosions, the *Flexotir* used only fifty grams of dynamite per shot, detonated inside an iron-perforated sphere about a dozen meters below the seawater surface. It was designed for what the oil industry called *"oceanography of great reconnaissance"*: a device intended to survey vast underwater regions by acquiring continuous data over kilometers while the vessel remained in motion.[35] In France, no other academic institution had comparable geophysical technology.

Operations for offshore oil exploration began immediately after *Flexotir*'s development, across the North Sea and on the ample and shallow continental shelf of France's North Atlantic coastline—the most promising regions for hydrocarbon production in the short term.[36] Almost nothing was known about the origin and geological structure of the French continental shelf: at that moment, it was a barely explored ground, whose economic promises were "completely unknown and unpredictable."[37]

With the *Flexotir*, French oil companies now had access to information about the seafloor unattainable by academic research teams. Meanwhile, geophysicists at French universities and research centers were leading the exploration of the Mediterranean seafloor using self-made or American-manufactured seismic sources: the *sparker*, which generated acoustic waves by releasing electric sparks, and the *boomer*, the electromagnetically driven seismic source designed by Harold Edgerton. Both of these devices produced high-frequency waves, which generated high-resolution data but at a low penetration rate. This meant that academic geologists could only study with accuracy the first dozen meters below the seabed, but they could not obtain any clues to the deeper layers.[38] Unlike these technologies, dynamite explosions in the *Flexotir* generated a low-frequency signal that could travel much deeper below the seabed and offer high-resolution data from deep sedimentary structures. Yet, the *Flexotir* was out of reach for academic researchers: although the mechanics and techniques of light seismics (like the *sparker* and *boomer*) could be easily transported from one vessel to another, the *Flexotir* weighted eight hundred kilograms and required a specific ancillary gear, including expert technicians, to operate it.

Other technological advances set the IFP at the forefront of seafloor exploration in France, not only in marine geophysics but also in marine geological methods as well. A new sampling system, the electrocorer, could retrieve one-meter-long cores of soft sediment from the seabed, even at depths of five hundred meters. Unlike other coring systems available at university departments, the electrocorer was designed to perform amid strong underwater currents. The coring system was lowered and suspended from the vessel through a flexible pipe—thus reducing the pressure exercised by underwater currents; and it stood firmly on the seabed while coring thanks to a solid, tripod-like base.[39] But its burdensome infrastructure and dependence on an electric source prevented its use by others than the oil industry. In the field of offshore drilling, the IFP's Division of Exploitation also made progress patenting the "Flexo-driller" in July 1965, a drilling system that owed its name to its unique flexible drill pipe. The entire system could be installed on vessels or barges, transforming them into drilling platforms. Its mobility and easy assembly were features devised to speed up boring operations over the continental shelf in the exploratory phase of oil production.[40]

During its first years of existence, CEPM greatly contributed to the exploration of the seafloor. Thanks to the oil industry's generous economic support, the IFP's vessels could work at sea almost year-round, acquiring a massive amount of new data with their own devices. At the IFP, geologists at the Division of Geology processed and interpreted the data, drawing the first geological maps of the seafloor.[41] Continental shelves in the North Sea, the Bay of Biscay, and the western Mediterranean, the most interesting regions for exploiting hydrocarbon deposits due to their shallowness, began to acquire their own contours, shapes, and character. However, the scientific expertise available within the oil industry soon felt short. If the aim was to speed up the detailed reconnaissance of vast underwater regions, new partnerships needed to be established.

SCIENTIFIC EXPERTISE MEETS INDUSTRIAL MARINE EXPLORATION

Under the leadership of André Giraud and the pooled resources of industrial stakeholders, France's oil industry was soon equipped with the latest technological advancements. Yet CEPM's interests immediately turned to academic research. According to Giraud, fundamental research was crucial for the nation because it helped achieve economically relevant goals. CO-MEXO, the expert committee striving to move forward France's oceanographic capabilities, could efficiently hinge between marine research and advancing the nation's economy.

Grasping the difference between academic and industrial seafloor exploration is key to understanding why the latter began requiring the former. Both groups coincided in their eagerness to sketch the seafloor's main sedimentological features, to correlate them with mainland formations, and to date their deposition; to spot cracks, faults, and fracture zones that could jeopardize future oil-exploitation activities or prove key geological-historical events; and to draw geological syntheses, like maps and geological profiles, from the incoming data. But the two clusters differed in their research approaches, which in turn were determined by the time, technologies, and resources obtainable. Petroleum geologists tended to move quickly from one region to another, investing large sums of money to survey each area of interest. These researchers had no chance to develop a personal scientific attachment to a particular region, geological formation, or historical event. To stay competitive internationally, when the oil firm considered that the knowledge acquired was sufficient, petroleum geologists were compelled to move to a different geographical area. The North and the South Atlantic Ocean, the Indian Ocean, the Persian Gulf, or the Mediterranean Sea—across all these regions, the IFP technologies were tested in barely five years. Conversely, academic geologists tended to specialize in particular geographical regions, chronological periods, and research lines, topics which young recruits frequently inherited from senior researchers. Limited research budgets, which constrained the days spent at sea, the number of oceanographic campaigns to be undertaken, and the research techniques available, impelled them to develop an exhaustive knowledge of the regions near their institutional bases. At the Oceanographic Observatory of Villefranche-sur-Mer, for instance, Jacques Bourcart left a crew of geologists focused on studying marine regressions in the western Mediterranean, while at Monaco, students collaborating with the team of Jacques Cousteau became proficient in the technical details for conducting seismic surveys across the Mediterranean. Along the Atlantic coastline, oceanographer Louis Dangeard had pioneered studying the region's sedimentology, training future marine geologists at the Universities of Brest and of Rennes. Scientific forums, where information was exchanged and discussed, enriched geologists' interpretations of the seafloor's geology, dynamics, and history.

This framework forged the academic community into a pool of geological knowledge and expertise on which the French oil industry could rely. Industrial managers did not need them to directly identify hydrocarbon deposits but rather to contribute their detailed knowledge of the seafloor's structure, dynamics, and composition. This information constituted the foundation on which general guidelines for hydrocarbon detection could be designed. For the oil industry itself to develop such expertise would have consumed an excessive amount of funds and time.

Reflecting the need for this expertise, petroleum experts gradually established informal, one-to-one relationships with academic researchers. An illustration for the North Atlantic seafloor: Gilbert Boillot, geologist at the University of Rennes and former student of Jacques Bourcart. In 1964 he concluded the first sedimentological study of the Breton continental shelf by applying the sampling and dredging methods learned from his master to map the distribution of sediments in those shallow, near-shore seawaters (a potentially productive area, if hydrocarbon deposits were identified). Boillot's work attracted the attention of Étienne Winnock, a petroleum geologist at the oil company National Society of Aquitanian Petroleum (SNPA) and responsible for drafting a preliminary report on the oil potential of that same region. To speed up his work, Winnock requested Boillot's contribution to the inventory, an offer that the geologist accepted as long as the oil company provided him a rock-coring instrument. Shortly after, the University of Rennes received a brand-new Stettson-Hill piston corer, an English-manufactured device not available at any other French laboratory. As Boillot later acknowledged, the economic and technological contributions of the oil industry to his modest laboratory transformed him into a "full-time marine geologist" who could study the geological history hidden under the seafloor beyond its sedimentary cover.[42] As for the Mediterranean seafloor, the team of marine geophysicists at the Oceanographic Museum of Monaco (led by Jacques Cousteau) soon after received a request from ERAP managers. Given the team's vast expertise in conducting seismic surveys, its scientists were asked to draft a bibliographical report, detailing the particularities of different sounding instruments and defining the most appropriate techniques for hydrocarbon exploration.[43]

As petroleum experts realized the cost-effectiveness of recruiting academic research teams for assisting in their underwater exploratory tasks, industry representatives sitting at CEPM agreed to establish formal venues of collaboration. Giraud, president of the committee, took charge of voicing the oil industry's needs, priorities, and concerns at the one national forum devoted to enhancing scientific ocean exploration: COMEXO. His influence proved crucial to transforming COMEXO's priorities from basic research to the exploration of natural resources, with the crucial backing of the Interministerial Committee and representatives of other oceanic industries.

A SPILL OF INDUSTRIAL PRIORITIES INTO ACADEMIC RESEARCH

The new orientation of the French oil industry toward the offshore drastically transformed national priorities over the oceans, influencing the government's perception of the activities that ought to be pursued there. In 1965, when COMEXO underwent its end-of-term performance review, a new governmental priority was explicitly articulated in favor of keeping up support. As in the initial evaluation five years earlier, delegates at the

DGRST, the funding administration, praised ocean sciences as a means to enhance France's international prestige; but they newly identified the exploitation of natural resources as a major driving force. They justified their decision to the Interministerial Committee by arguing, "Today, France cannot ignore the oceanic space that surrounds it, and which other modern nations seek to explore and are getting ready to exploit. [. . .] The seas are not anymore simply means of communication, for which to ensure the control of its surface. They offer an environment where scientific adventures will precede a rational exploitation of the wealth contained in the water mass or in the seafloor."[44]

In contrast to the rationale revealed five years before, delegates at the DGRST at this moment poetically acknowledged the ocean's potential beyond its surface, a new perception of the seafloor's increased importance for its hidden resources. In this view, the seafloor was recognized as a three-dimensional space whose hidden and barely reachable natural resources needed to be in the spotlight of scientific and technological efforts. Scientific research was no longer presented as an activity pursued for the sake of knowing the marine environment but instead portrayed as an "adventure" devoted to supplying the nation's needs in terms of food and economic resources while keeping pace with other nations.

The requirements of the new offshore industry defined this new stance but were not the only motives. Recent oceanic advancements in the US also became powerful driving forces and a model for France's oceanic reorientation. American President Lyndon B. Johnson, in office from 1963 to 1969, was excited about the future possibilities of marine resources for his country and pushed for huge national efforts to promote oceanography.[45] In Johnson's view, ocean sciences were a suitable means to ease political tensions through cooperation, as well as a possible solution for solving globally pressing problems of resource scarcity. In a 1963 address to the United Nations, Johnson signaled ocean exploration as a future site to build international relations and exert diplomatic efforts.[46] In that single year, the US spent ten times more for oceanography than France had invested over the five-year period 1961–66. In 1965, the US Congress designed a national plan to drive oceanographic research toward the fulfillment of national needs, which crystallized in the creation of the Commission on Marine Science, Engineering, and Resources. This commission would establish a foundation for what was expected to become a "NASA of the oceans," an institution that would implement and coordinate a national oceanographic program.[47]

This situation sparked anxieties among the Interministerial Committee. André Giraud, spokesperson for the national oil industry, openly expressed his fears to the minister of scientific research, Alain Peyrefitte, stating in a letter that the US was "preparing to launch an offensive to discover all the riches of the seas."[48] A chorus of other ocean enterprises, such as institutions for fisheries, the French Navy, and the merchant marine, joined the voice of extractive industries, expressing equivalent concerns and supporting the integration of scientific research with all sorts of oceanic industries. International competition was a strong motive for increasing investment, yet the fear of losing a propitious place in future oceanic industries also proved a firm reason to push for stronger, national control over ocean exploration. Consistent with this argument, the DGRST explicitly alluded to the importance of positioning France at the forefront of the just-launched international competition in the oceans in its justification for enduring support of COMEXO.

In this framework, the oceans flowed to the core of France's economy. The *valorisation des océans*, which roughly translates as "enhancement of the oceans" or "optimizing the ocean's value," became a political motto to express the overarching aim of this new oceanographic policy.[49] It is worth pausing here for a moment to consider the term *valorisation*, whose singularity and nuance are distinctive. Although difficult to accurately translate in English, the term is commonly used in French. The Larousse dictionary defines it as "an increase in the market value of a product or service, caused by voluntary maneuvers or a legal measure" or as "the action of giving more value to something or someone"; whereas in a broader context, economic Marxism defines the "valorization of capital" as an increase in value of capital assets through value-forming labor during the production phase. This is to say, "to valorize" normally implies a relation to an economic value. In the sixties and seventies, the term *valorisation* became a ubiquitous companion to France's scientific and industrial policies—for example, the National Agency for the *Valorization* of Research was created in 1967 to *valorize* (or obtain larger economic revenues from) scientific research, by exchanging technologies and sharing results with industrial organizations. But, regardless of any widespread use, words are rarely neutral.[50] By using the word *valorisation*, the French government was making explicit its renewed expectations about the oceans' submerged territory: that the investment devoted to their exploration would be returned in the shape of valuable resources and economic revenue. That is, producing knowledge for the mere sake of knowing was no longer sufficient. Instead, technoscientific innovations were required to yield useful products: raw materials, food, or energy supplies. In other words, the valorization of the oceans implied that the economic revenue produced from exploring the oceans would be expected to surpass the investment devoted to scientific research, technological innovation, and equipment.

Starting in 1966, COMEXO activities were expanded for one more year with one more member: Giraud became vocal among marine scientists and other oceanic experts, a position that symbolized the new union between the oil industry and academia. The plan was to ensure a smooth transition from this scientific committee to the politically led structure for ocean exploration that was being discussed in parallel at a ministerial level. Taking stock of the committee's past achievements and weaknesses, COMEXO had managed to build the R/V *Jean Charcot*, the first oceanographic vessel in France able to conduct interdisciplinary surveys on the high seas, and its oceanographers, biologists, and geologists anticipated a future organization for oceanography.[51] However, they also acknowledged their inability to establish an efficient training plan and admitted that research remained dispersed across more than a hundred laboratories.[52]

During the second meeting of the renewed COMEXO, Giraud sketched the principles of the new oceanographic policy decided at the Interministerial Committtee. After an opening speech pervaded with an overtone of expansion, occupation, and conquest of the oceans, Giraud described the measures embraced to promote oceanography. Decisions would be taken at an interministry level, specific research equipment and laboratories would be renovated, and links between "basic" research and the "valorization" of the oceans would be strengthened. Only research lines aimed at achieving particular economic goals were to be supported.[53] These actions would be coordinated under a centralized structure, but whether it would be a brand-new national oceanographic center or another kind of government structure was still under official discussion. What was clear, though, is that the ambitions France projected onto the new center reached beyond its national economy and territorial control. The institution was also called upon to perform a diplomatic role and to restore France's international prestige by performing cutting-edge marine technoscientific research. Members at the Interministerial Committee expected it to become a key player in organizing European oceanography, and to constitute a counterbalancing oceanographic force between the US and the USSR.[54]

This approach to near-future research was neither criticized nor contested by academic researchers sitting on the committee (at least, not in official documents), probably because scientific interests were aligned with industrial ambitions.[55] The measures Giraud presented were in principle beneficial for academic researchers and their institutions: increased funding for marine research, some freedom to continue allocating the funds, and the creation of an institution to bolster France's position in international science.[56] Basic research was not (apparently) at risk, since almost any research line could provide knowledge useful for both basic research and to support the exploitation of natural resources. In marine geosciences, for instance, better knowledge of sedimentation processes could help identify areas were hydrocarbons, minerals, and other useful materials could accrete. Designing topographic and geological maps of the seafloor was as useful for scientific research as it was for exploiting deposits of minerals and hydrocarbons.[57] Moreover, the new policy would contribute to institutionalizing exchanges between academic researchers and the offshore oil industry, to mutual benefit.

OCEAN SCIENCES, INDUSTRY, AND POLITICS CONVERGE

France's new ocean policy was explicitly inspired by the American approach. The Interministerial Committee and André Giraud supported the creation of a permanent agency, similar to the proposed "wet NASA" or "NASA of the oceans"—which the US eventually established in 1970 as the National Oceanic and Atmospheric Administration (NOAA). In France, the key question was whether the new institution should centralize authority over national ocean policy. The structure of this institution was also a subject of debate. Giraud, for instance, opposed creating a research center that combined the coordination of research in French coastal and international waters, the development of academic-industrial partnerships, and efforts to match American oceanographic capabilities.[58] Other observers argued that a government agency would be the most effective structure to blend political and economic goals with scientific research.[59]

In November 1965, representatives from various ministries convened at the DGRST offices to weigh whether the best strategy was to establish an organization within the public administration or to create a national oceanographic center (not a *research* center), modeled after the French National Center of Space Studies (CNES, inaugurated in 1961).[60] The first option involved creating a council, led by a minister or state secretary, to coordinate research across various ministries and institutions. It was soon discarded because coordination would be impossible if ministries were not willing to cooperate. In contrast, CNES was an example of successfully coordinating efforts under the Ministry of Defense, the Ministry of National Education, and the prime minister's cabinet. Decisions favored the establishment of a national center. Industry experts would occupy decision-making seats to articulate their needs with the research promoted; yet the center's multiministerial directive board would retain control over the budget and responsibility for managing national resources such as vessels, submersibles, and researchers.[61] However, disagreements persisted regarding the scope of the center's responsibilities—for instance, whether it should include military research, oversight of nuclear tests in the Pacific, or hydrocarbon exploitation—as well as which ministries and priorities should be involved, who would hold decision-making authority, and how it would relate to the Interministerial Committee.[62]

The final decision was made on April 22, 1966, when the Interministerial Committee issued an official report announcing the creation of a new institution to coordinate oceanographic research: the National Center for the Exploitation of the Oceans, or CNEXO.[63] Giraud retained his influence, taking on the role of France's oil industry representative within the new institution, which was tasked with overseeing national ocean exploration and future resource exploitation. In this way, the connection between the French oil industry and academic marine geosciences was expressly institutionalized.

Just after issuing the official order instituting a national oceanographic center, the minister for scientific research and atomic and space affairs, Alain Peyrefitte, called scientific adviser Yves la Prairie to draft the bylaws for the new center in coordination with the involved ministries (to specify its regulations, goals, and detailed mechanisms of articulation with national and private institutions).[64] La Prairie was not an oceanographer or a researcher, and he had never had any contact with the marine environment beyond his former experience as an official of the French Navy. He was a civil servant, known among the ministerial circles because "he liked things about the sea" and for his support of the Gaullist government.[65] Yet he would become a key figure in France's ocean exploration throughout the 1970s.

Trained as a naval officer in Toulon, Yves la Prairie spent the first half of his career as a frigate commander, sailing the Mediterranean across North Africa and the Middle East. That period reinforced his nationalist, pro-Gaullist ideology. He was concerned about the consequences of the nation's loss of control over colonial territories and wished to see France's prestige restored.[66] In 1954, la Prairie set his military career aside to settle with his family in Paris. Through his contact with other military commanders and engineers relocated to civilian positions, he obtained a position at the French Atomic Energy Commission (*Commissariat à l'Energie Atomique,* or CEA), which turned into a relevant background for his future role in ocean exploration. Created in October 1945 by General Charles de Gaulle, the CEA was designed to exploit the scientific, industrial, and military potential of atomic energy, becoming the materialization of the Gaullist ideology consistent with restoring France's *grandeur* through technological prowess.[67] La Prairie was appointed secretary to Jacques Yvon, director of the Department of Atomic Reactors, where he developed firsthand knowledge about the workings of big science institutions, became skilled in interacting with high officials, scientists, and policymakers, and familiar with the political apparatus through which scientific policies materialized in research projects.[68] In the following years, he moved from the CEA to rise to technical counselor to Gaston Palewski, minister of scientific research and atomic and space affairs. In that position, la Prairie learned how decision-making was implemented at the highest political levels.[69]

La Prairie's biographical context is important for understanding why Peyrefitte chose him to draft the bylaws of the future oceanographic center. The ocean was emerging as a technoscientific field, as relevant as space and nuclear research, through which France could increase its international recognition and national development. In comparison with nuclear research, exploring the oceans would require a national structure equivalent to the CEA, which could drive technological innovation and scientific research toward the fulfillment of industrial, diplomatic, economic, and military interests. Alongside nuclear research, advancements in ocean exploration symbolized the nation's technological prowess. Yet, in terms of the materiality of exploration, knowing the ocean was more like space research: both were undiscovered, infinite, and almost intangible territories, and both required large economic investments and a dedicated human force to be explored. While by the mid-sixties French government officers assumed the oceans to be of "lesser political interest" than outer space, their immediate economic value was recognized as much higher.[70] In this framework, it is no wonder that the nature, goals, and structure of the new oceanographic center were explicitly inspired by the two national centers devoted to nuclear and space research, the CEA and CNES.

The National Center for the Exploitation of the Oceans (CNEXO) was inaugurated in April 1967. It was not a scientific research center but a politically driven institution focused on implementing a national oceanographic policy, supervising its development, steering marine sciences toward industrial goals, and ensuring French participation in international ocean research. Its overarching objective was to increase the efficiency of offshore exploration by coordinating research efforts with various marine industries. By fostering collaboration between the offshore oil industry, national ministries, and academic institutions, CNEXO sought to orchestrate oceanographic programs with a dual agenda: advancing scientific understanding of the oceans while exploring their economic potential. For marine geosciences, this meant advancing knowledge of the ocean crust's dynamics and composition while supplying crucial information about hydrocarbon and mineral resources.

Like the CEA and CNES, the national coordinating agencies in nuclear and space research, CNEXO held the legal status of an industrial and commercial establishment under the prime minister's administration, which granted its managers absolute autonomy in administering their budgets while freeing them from dependence on other ministries. Selected representatives from public institutions and private companies composed its Administrative Council, which facilitated the fulfillment of targeted economic motivations through scientific research. Jean Cahen-Salvador, state councillor and former head of the French Aerospace Industries Association, was appointed first president of CNEXO's Administrative Council. Under his leadership, twelve representatives from the ministries of foreign relations, finances, higher education, fisheries, transportation, industry, and the army voiced their priorities to define CNEXO's research agenda. Giraud, representing the French oil industry, ensured the connection between industrial goals and CNEXO's future projects. Oceanographers were relegated to the role of technical advisors at CNEXO's Scientific and Technical Committee. Most of its members had been part of the former COMEXO, including oceanographer Henri Lacombe but, unlike in the previous committee, the decision-making capacity of this group was demoted and placed under CNEXO's general director, who would play a mediating role between oceanographers and policymakers at the Administrative Council.

The position of general director was granted to Yves la Prairie because, as the new Minister of Scientific Research Alain Peyrefitte asserted, CNEXO needed a head who "was not an admiral, nor a naval engineer, but someone younger and external to the military. . . . neither a university scientist, given CNEXO's orientation toward applied and economic goals . . . nor an expert from private industry, at least at the beginning."[71] La Prairie's complete lack of oceanographic training was irrelevant: the Scientific and Technical Committee would stand at his side, providing him the scientific and technical advice he would need.[72] More important was his civil service background, mixed enough to mediate between political, military, scientific, and international interests. In the eyes of skeptics, he was no longer a military officer, nor an expert from the oil industry. He was a policymaker, someone who would be capable of channeling the ideas and priorities from the ministries, putting state interests ahead of scientific ones.

FROM COLONIAL TERRITORIES TO THE
INSTITUTIONALIZATION OF OCEANIC AMBITIONS

The CNEXO's inauguration completed a process that redefined the seafloor: in France, the seafloor was now (politically) perceived as an underwater territorial extension. This new perspective mirrored onto the oceans the strategies, activities, and economic ambitions once directed toward former colonial territories.

The instability of France's hydrocarbon suppliers played a key role in driving this new oceanic perception, first among oil industry managers and later within government circles. As a result, the ocean strategy France developed for the seafloor was influenced by a colonial perspective, focused on achieving energy independence and expanding global influence. Leading civil servants like André Giraud viewed the oceans as "virgin territories" ripe for conquest and exploitation, reflecting ideological continuities from the mainland to the offshore.

With the seafloor now seen as a promising frontier for securing strategic resources such as oil, gas, and minerals, the anticipated industrial exploration increasingly called for marine geosciences. A formalized governmental effort soon replaced initial attempts at academic-industrial collaboration. The thriving period of seafloor exploration that ensued was deeply shaped by the close collaboration forged between public officials, industry leaders, and academic institutions. CNEXO became the hinge that bridged these domains and aligned their efforts. The establishment of this institution signaled a strategic pivot in France's national priorities, blending scientific exploration with economic imperatives, and marking a new chapter in the nation's approach to securing natural resources.

CHAPTER 4

Three-Dimensional Territories: Science and Industry in the North Atlantic

But why drives on that ship so fast,
Without or wave or wind?
The air is cut away before,
And closes from behind.
Fly, brother, fly! more high, more high!
Or we shall be belated:
For slow and slow that ship will go,
When the Mariner's trance is abated.

SAMUEL TAYLOR COLERIDGE, *THE RIME OF*
THE ANCIENT MARINER

In April 1968, *Le Monde* published an article entitled, "An Attempt at
Research-Industry Cooperation: The Next Cruise of the *Jean Charcot*."[1] The
newspaper introduced an unprecedented French oceanographic survey in
the Gulf of Guinea, a region located off the western North African coast.
Initially envisioned as a cruise focused on marine biology, the sudden interest in the region by the French petroleum consortium CEPM transformed
it into a multipurpose mission. Now, geologists and biologists, academic
researchers and industry experts would share floating space onboard the
R/V *Jean Charcot* while dividing the time at sea between marine geophysical surveying and biological sampling. The oceanographic vessel turned
into a place where two research rationales, the scientific and the industrial,
converged in a joint mission.

CNEXO's leadership made possible such an integration of industrial
interests within scientific expeditions. The institution became a key vector to deploy France's political agenda in the oceans (through sciences):
marine geophysical surveys served to stress the country's presence across
the oceans, introduced French researchers into the study of plate tectonics
theory, and enhanced resource assessment at home and abroad. This all-
in-one blending in seafloor exploration reflected the emergent imaginary

of a global future tilted toward the oceans, in which technosciences would serve state oceanic interests. Metaphorically, seafloor exploration in the early seventies mirrored Coleridge's *Rime of the Ancient Mariner*. Just as the Mariner's ship is driven by uncontrollable forces beyond human understanding, CNEXO's oceanographic vessels were propelled by another kind of force—political and industrial imperatives, where national interests and economic goals fueled the exploration. The Mariner's ship slows as his trance subsides; similarly, the frenzy of large governmental investments in seafloor exploration would only subside once economic realities set in.

This chapter focuses on both the contribution of the French oil industry to the oceans' geological knowledge and the contribution of academic geology to resource exploration. Both developments progressed inseparably. This idea unfolds through two research activities across the North Atlantic between 1968 and 1971. The first was a series of hybrid cruises devoted to exploring both tectonic dynamics beneath the high seas and the economic potential of continental margins. The second was a coordinated national effort to map France's continental shelf while unraveling its geological history. In both, scientific interests intermingled with France's industrial expectations and geopolitical priorities.

DRAWING LINES OFF GUINEA: SEISMIC PROFILES, TRANSFORM FAULTS, AND OIL HORIZONS

From the first meeting of CNEXO's Administrative Council, Yves la Prairie and André Giraud established conversations to negotiate the question that had stimulated the institution's creation: how to formalize industry-academia relationships.[2] Contributing industrial research resources (experts, vessels, funding, and research technologies) to joint scientific surveys appeared a canny strategy to cut down costs and speed up the industrial reconnaissance of the seafloor, and would enable academic researchers to work with industrial-quality data. There was yet another motive to foster such ties: Under CNEXO, scientific expeditions could facilitate the industrial exploration of seawaters beyond France.

At the heart of this motivation lay the Law of the Sea's framework, which seemed to offer flexibility to merge scientific campaigns with preliminary surveys of industrial interest. The 1958 Convention on the Continental Shelf—approved during Geneva's UNCLOS I and implemented in 1965—had stimulated coastal governments to define a strategy to assess the seafloor's economic potential. This legal framework not only delineated the boundaries of the submerged national territories (200 meters depth *or* the limits of technological exploitation) but also established rules for hydrocarbon and mineral exploration in foreign continental shelves. Coastal governments required public and private companies to obtain exploration leases, which set the terms and conditions for operations—much like the regulations governing onshore hydrocarbon exploration.[3] In contrast, regulations for scientific research were more flexible. While researchers needed to seek permission from the coastal government before launching an expedition, the Convention on the Continental Shelf mandated that the coastal state "shall not normally withhold its consent if the request is submitted by a qualified institution with a view to purely scientific research into the physical or biological characteristics of the continental shelf."[4] Under these provisions, marine geophysical campaigns under CNEXO's sponsorship could benefit French oil companies by sharing data collected during scientific missions to identify regions with potential oil reserves.

Negotiations between la Prairie, as CNEXO's director, and Giraud, as spokesperson of the oil industry, resulted in April 1968 in the first industrial-academic campaign, Guinée I, along the shores of Ivory Coast and Guinea. *Le Monde* announced with fanfare the hybrid nature of the mission, underscoring both the distinct turn of marine sciences toward economic revenues under CNEXO's guidance and the promising future this relationship held for ocean research.[5] The mission was initially conceived by marine biologists at the Office of Scientific and Technical Overseas Research to identify promising fisheries for yellowfin tuna onboard the brand-new *Jean Charcot*. However, as the region drew the interest of oil firms, the goals of the mission changed. In February 1968 the IFP and the oil company ERAP requested CNEXO, as the mediating institution, to include its geologists and oil engineers on the voyage to assist with studying the geological structure of the region.[6] In exchange, CEPM contributed a "fee" of 611,000 US dollars and, perhaps most importantly, the seismic source *Flexotir* with two technicians.[7] Installing such an industrial technology onboard an emblematic oceanographic vessel was laden with symbolism: it implied equipping the only large oceanographic vessel in France, designed to conduct scientific missions, with a device built for offshore oil prospecting.[8]

On May 20, 1968, a team of two academic geophysicists and four petroleum geologists boarded the *Jean Charcot* at the port of Abidjan.[9] Using marine geophysical methods, the team investigated fracture zones and thick sediment accumulations across the Gulf of Guinea, spanning from the continental shelf to the abyssal plain.[10] In their mission, scientific and industrial interests converged. The findings aimed to identify sedimentary regions suitable for hydrocarbon deposits, particularly beneath the shallow continental shelf, while advancing France's role in the burgeoning field of seafloor spreading research through the study of transform faults in deeper areas.[11]

The concept of transform faults had recently been introduced by Canadian geologist J. Tuzo Wilson. Wilson sought to explain the abrupt disruptions in seismic activity along mid-ocean ridges, a phenomenon he observed while studying arc islands in the Pacific Ocean. He proposed that Earth's surface consists of rigid plates separated by a belt of weaker regions forming mid-ocean ridges, mountain chains, or major faults with significant horizontal motion. Transform faults, he argued, serve as junctions connecting these features, where plates slide past each other in opposite directions. This concept was pivotal in bolstering the hypothesis of seafloor spreading proposed in 1962 by American oceanographers Harry H. Hess and Robert S. Dietz, as transform faults could only exist if the crust was indeed moving.[12]

In 1967, Lynn Sykes, an American seismologist from Lamont, validated Wilson's hypothesis through global seismological data from mid-ocean ridges.[13] These results resolved doubts that the ocean floors were splitting apart. By 1968, plate tectonics emerged as a comprehensive theory, synthesizing data and theories from American and British institutions. Working independently, Daniel P. McKenzie and Robert L. Parker at Scripps Institution of Oceanography, along with Jason Morgan at Princeton University, formulated the plate tectonic model. Meanwhile, French geophysicist Xavier le Pichon, at Lamont, integrated these data into a world map and calculated plate movements using paleomagnetic records.[14] Plate tectonics, the unifying theory of earth sciences, had just been born.[15]

The CNEXO expedition in the Gulf of Guinea marked the entry of French geoscientists into the forefront of the plate tectonics debate. Back on the mainland and after processing, interpreting, and discussing the new data, results were published in the international journal *Earth and Planetary Science*. The publication combined those results with new geological insights from the African continental margin, acquired during oil surveys. It was a reevaluation of the seafloor's structure in light of continental drift, which partially relied on industrial-quality data.[16] Because sediments in transform faults accumulate to great thicknesses, it was essential to use high-penetration geophysical devices—available only in the industrial sector. Using *Flexotir* seismic profiles, the French geophysicists identified deep fracture zones between the equatorial Atlantic and the Gulf of Guinea. These were evidence of the ocean's expansion, driven by the movements of transform faults.

Alongside its geological breakthroughs, the Guinée I campaign demonstrated that close collaboration between the oil industry and academic researchers was not only possible but productive for both. The oil industry could leverage scientific expeditions to conduct preliminary explorations of areas of industrial interest, while academic researchers benefited from access to cutting-edge geophysical technologies and industrial-quality seismic profiles. When disseminated through academic publications, the industrial data became a crucial contribution to the scientific understanding of the ocean's crust.

The campaign's success spurred la Prairie to negotiate with Jean-Claude Balanceanu, the general director of the IFP, to install a permanent *Flexotir* system onboard the *Jean Charcot*. This effort raised unprecedented challenges. La Prairie and Balanceanu needed to establish clear boundaries between basic and industrial research to prevent overlapping interests and the misuse of data.[17] They developed an operational framework tailored to each type of research and crafted a secrecy regime that struck a delicate balance: loose enough so geologists and geophysicists could publish their findings, while ensuring that the oil industry retained some control over the data. The contract, finalized in July 1969, defined the *Flexotir*'s use for scientific research. The IFP agreed to provide the technical expertise, savoir faire, and operational support needed to deploy the device, while CNEXO committed to sharing all geophysical data with the IFP. The contract also imposed an eighteen-month moratorium on publishing data after each campaign.[18] This arrangement granted oil companies a precious time window during which they had exclusive rights to use geophysical data for their own goals, before it was disseminated across academic domains.

After the agreement's signature, CNEXO began to organize geophysical surveys utilizing the *Flexotir* onboard the *Jean Charcot*. Some of those campaigns were genuine international displays of France's newly acquired oceanographic capabilities, in which the *Jean Charcot*, research technologies, and numerous researchers from different backgrounds were mobilized to undertake large-scale oceanographic missions, backed by generous economic support. Campaign Noratlante I, conducted in 1969, clearly illustrates this mode.

VENTURING INTO THE NORTH ATLANTIC:
RIFT DYNAMICS AND OIL RIMS

The scientific cruise Noratlante I (a name blending "North" and "Atlantic"), across the mid-ocean ridge, was primarily aimed at introducing France into the international research stream of plate tectonics. Despite its chiefly scientific nature, the cruise was as useful for understanding Earth's dynamics as for achieving CNEXO's implicit motivations—to stress France's presence on the high seas and to assess the economic potential of continental margins around the world.[19]

Marine geophysicist Xavier le Pichon, organizer of the cruise and key contributor to plate tectonics theory, embodied CNEXO's desired international trajectory. Le Pichon, born in 1937 in the Vietnamese region of French Indochina, had spent most of his career in marine geophysics at Lamont. After graduating in geophysical engineering at the University of Strasbourg, in France, le Pichon obtained a Fulbright Fellowship that brought him to the US. His plan was to teach and conduct research on geophysics at Columbia University, but destiny led him to meet Maurice Ewing and his team, by then focused on conducting marine research to prove seafloor spreading. Under Ewing's mentorship, le Pichon began his career as a promising marine geoscientist surveying the mid-Atlantic and Pacific Ocean ridges. By 1968 le Pichon had gained international prestige in the geological community with his publication of a quantitative model of plate tectonics theory—fame that immediately caught la Prairie's attention.[20] He wanted le Pichon to lead France's marine geosciences, both for his vast experience in the American research system and his international reputation, which would build up CNEXO's international prestige (fig. 11).[21]

Figure 11. Xavier le Pichon (right) in conversation with American oceanographer Walter Munk in 1972. While affiliated with CNEXO, le Pichon promoted and maintained collaborations with American marine researchers, becoming an ambassador of France's marine geosciences abroad. Source: Niels Bohr Library and Archives, American Institute of Physics. Reprinted with permission of the American Institute of Physics.

In January 1968, la Prairie offered le Pichon a permanent senior position at the newly created CNEXO.[22] Le Pichon hesitated to accept it—he valued the bright professional future that US laboratories offered to him and the opportunity to focus on conducting research; yet he accepted the advisory position la Prairie offered on the promise of organizing and leading his own scientific cruises on the high seas. One month later, le Pichon joined as scientific advisor to the general director, responsible for creating a team of marine geoscientists. He suddenly found himself drowned by administrative responsibilities, endless meetings, and uninterrupted report-writing, but these difficulties were compensated by his ability to recruit a research team, an exciting task shaped by his own views.[23] After recruiting a handful of geologists scattered around university departments and marine stations, le Pichon organized the Noratlante I cruise that, in addition to pursuing scientific goals, was to instruct them on the latest theories, concepts, and research methods for studying plate tectonics.[24] With le Pichon among its leadership, CNEXO became a key vector for channeling scientific knowledge from foreign institutions to France.

On August 1969, the *Jean Charcot* set sail from Brest with more than forty experts onboard—marine geologists, geophysicists, biologists, engineers, and technicians. In addition to CNEXO's young workforce, invited researchers from academic institutions in France and abroad embarked, while two engineers from the IFP ensured the proper functioning of the *Flexotir* throughout the three months of the voyage.[25] This interdisciplinary campaign on the high seas also made headlines in *Le Monde*, where it was presented as the most ambitious, largest campaign ever undertaken by the *Jean Charcot*.[26] Noratlante I studied rifting processes and transform faults across the North Atlantic continental margins by employing seismic, magnetic, and gravimetric techniques, a combination of research methods that, only a few years before, would have been inaccessible to academic marine geologists. Through magnetic techniques, the team measured the strength and direction of Earth's magnetic field, because magnetic anomalies—changes in the field's intensity or direction—indicated the presence of particular minerals and formations. Gravimetric techniques, measuring the gravitational pull of Earth, detected the presence of dense materials in the lower crust and upper mantle. Coupling these with seismic techniques, which provided information on the seafloor's layered composition, the team could investigate the ocean crust's composition, identify areas where the plates were moving in relation to one another, and obtain insights into the way Earth's crust and the upper mantle interact.

Scientific results, published in prestigious academic journals, engaged CNEXO's team in plate tectonics research. In *Nature*, for instance,

Noratlante I participants presented a model of the North Atlantic's evolution since the Cretaceous (140 to 180 MYA), when the North American and African plates began to drift apart.[27] As key evidence, the team had identified a continuous salt layer off Labrador and Newfoundland that correlated in age with salt deposits off Morocco, Portugal, and the Bay of Biscay. According to them, its formation dated from the initial stage of the North Atlantic's rifting process.[28]

A parallel reading of this cruise points to the role it was called to play in ocean geopolitics. By organizing a three-month-long campaign in the rough northern seas, onboard a brand-new oceanographic vessel that deployed industrial-quality technologies, France was displaying unprecedented technopower on the oceans. Trailing the persistent comparison with the US's oceanographic capabilities and activities, Noratlante I demonstrated a French presence in the world oceans via oceanographic research equivalent to the American. The academic publications on rifting processes beneath the high seas indicated that French geoscientists had entered a research field hitherto dominated by American and English researchers.

Although the cruise had not received any economic support from the oil industry, French oil companies could benefit from the Noratlante I campaign through the *Flexotir*'s deployment in scientific missions. As agreed between la Prairie and Balanceanu, geophysical profiles acquired during the campaign were readily available to the IFP and, by virtue of established relationships, to national oil companies. The mission provided relevant information on the economic prospects of the chief off-shore oil-harboring regions, the continental margins, a term embracing the continental shelf, rise, and slope. Even though their exploitation was not yet feasible, both technologically and because of legal uncertainties surrounding international waters, France's oil companies were interested in enhancing their geological understanding of the formation, dynamics, and age of continental margins, to assess their future economic potential.[29] This industrial facet was reflected in the delay of scientific publications that diligently respected the eighteen-month moratorium: CNEXO published a detailed volume including geophysical profiles along with magnetic and bathymetric data in February 1971, exactly eighteen months after the survey ended.[30]

For one more reason, the cruise Noratlante I was relevant to France's oil industry. Its scientific results piqued CEPM's interest in probing the economic prospects of the North Atlantic's continental margins, while confirming the value of relying on scientific teams for acquiring data. In January 1970, Xavier le Pichon again boarded the *Jean Charcot* as chief scientist of a hybrid scientific-industrial mission. The cruise Nestlante I (a contraction, this time, of "Northeast" and "Atlantic"), explicitly framed in the CNEXO-CEPM's cooperative program, was aimed at deepening the understanding of potentially productive areas already sounded during Noratlante I, such as the area off Morocco and southern Portugal, and the Bay of Biscay's northern region (fig. 12).[31] This time, though, the outcomes of the industrial-academic collaboration were mixed. In the cruise's final report, le Pichon displayed his dissatisfaction: The oil industry retracted its financial backing at the last moment, and IFP experts had to postpone their participation due to a simultaneous Mediterranean survey. But despite the apparent disappointment of the scientists involved, the geophysical profiles they acquired with the *Flexotir* flowed to France's oil companies by virtue of the technology's operating regime.

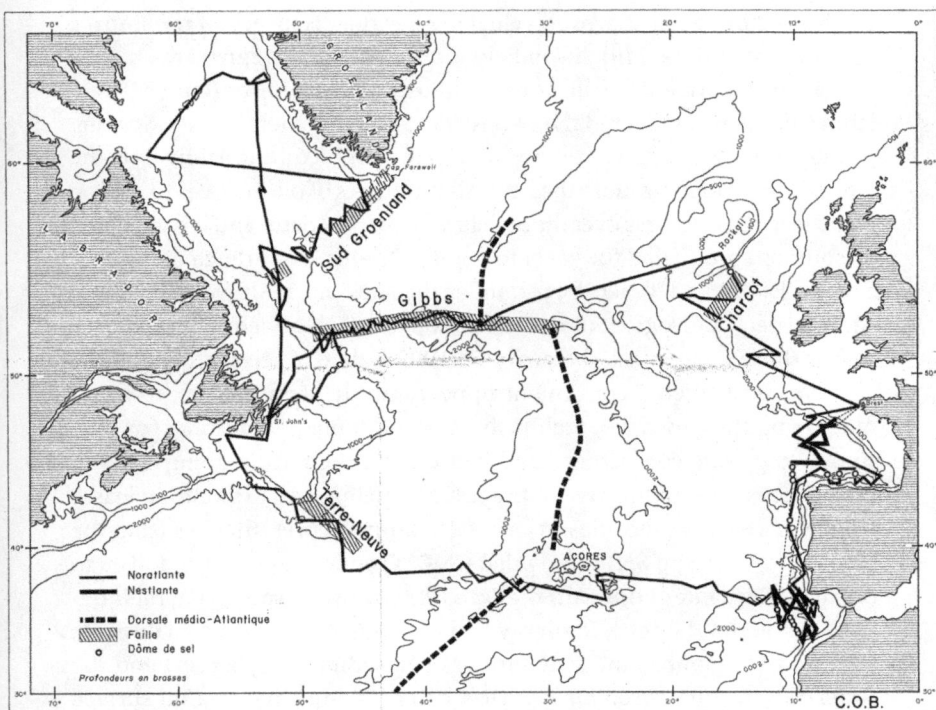

Figure 12. Map depicting the route of CNEXO's cruises Noratlante I (1969) and Nestlante I (1970), both across the North Atlantic Ocean. Source: Groupe scientifique du COB, *Résultats des campagnes du N.O. Jean CHARCOT*. Reprinted with permission of IFREMER.

MORE INDUSTRY THAN SCIENCE ACROSS
THE NORWEGIAN SEAFLOOR

Other oceanographic campaigns under CNEXO's auspices overtly constituted preliminary surveys for oil exploration, utilizing the *Jean Charcot* and human resources from CNEXO. The cruise Nestlante II, a follow-up mission to Nestlante I but further north, constitutes an excellent example of how predominantly industrial surveys contributed to foster the scientific understanding of the seafloor.

French oil companies chose to request CNEXO's support to survey the Norwegian seabed, a region of burgeoning economic interest. For the past decade, the North Sea had progressively become a key exploration ground for numerous European and American oil firms. The Groningen gas field, located in the Dutch sector of the North Sea, had become the largest and most productive gas field in Europe after production started in 1963. The British sector had also proved valuable after the discovery of the Betrice field in 1965 and the Murchison field in 1969. The Norwegian area turned into a benchmark for the offshore oil industry in 1969, after Philips Petroleum Company discovered the Ekofisk field at eighty meters deep. Starting production in 1971, Ekofisk became one of the largest oil and gas fields in the North Sea, producing more than 2.5 billion barrels of oil and over 10 trillion cubic feet of natural gas over more than fifty years to date (and it is expected to continue producing for years to come). In 1965, the French firms CFP and Elf-ERAP had joined forces under the subsidiary Elf-Norge to share surveys, data, and logistical resources to operate in Norwegian seawaters.[32]

For those companies, the novel CNEXO policy of scientific-industrial surveys constituted a convenient opportunity to reduce costs, to avoid mobilizing their own research resources and human workforce (or outsource to private companies), and it also freed them from going through the burdensome administrative process of obtaining oil exploration leases from Norwegian authorities. As a result, in 1970 representatives from the Elf-ERAP, CFP, and SNPA provided CNEXO almost 30,000 US dollars to organize Nestlante II onboard the *Jean Charcot*, while specifying their desired research. The seismic survey had to acquire data with two devices, a *Flexotir* (to glimpse the structure of deep sedimentary layers) and the American-manufactured *air gun* (to obtain high-quality data on surface sediments), as well as to deploy fifty sounding-buoys for seismic refraction to grasp the structure of the deepest oceanic crust.[33] The studied region was further north than the North Sea's oil-producing area, yet learning about the formation and sedimentary structure of the North Atlantic continental margin would enhance understanding of the commercial possibilities of deeper regions, potentially productive in the near future.[34]

The industrial bias of this oceanographic cruise was not a secret; rather, it emerged as a beneficial strategy for both French oil companies and Norwegian authorities. For the latter, allowing foreign scientific surveys with industrial relevance over their submerged territory granted new data and expertise. A Norwegian observer was invited to join the Nestlante II campaign, and all the data acquired was shared with industrial and academic institutions there.[35] As for French oil companies, yielding the mission's exclusiveness by labeling it as scientific survey—undertaken by a scientific institution, with a team of nonindustrial experts onboard—reduced the costs of an industrial reconnaissance survey, because the relevant international legislation permitted scientific research, not industrial prospecting. Since the recently enacted Law of the Sea compelled nations to acquire exploration leases from foreign authorities if aiming to conduct industrial exploration, articulating mission goals that were scientific meant that only the prior consent of foreign authorities (which should not normally be withhold) was required.

Following the same pattern as the previous hybrid cruises, Nestlante II provided crucial scientific information to describe the Norwegian continental margin in terms of plate tectonics. Geophysical data shed light on the opening of the Atlantic rift from the north of Iceland, on the processes that formed the Barents Sea, and on the contact zone between the oceanic and continental crusts from Spitsbergen to the Faroe Islands.[36] The French team published detailed accounts on the processes that opened the Norwegian Sea, characterized the formation of fracture zones in the Spitsbergen area, and demonstrated the sedimentary origin of the Vøring Plateau, a tongue-like structure that extends from the continental shelf into the deep oceanic domain.[37] Adhering to the guidelines of publication and disclosure specified for Noratlante I, the data did not belong exclusively to CNEXO but equally to the three oil company patrons. If data were of direct oil interest, the companies could reserve the right to prevent their publication.[38]

The marine geophysical campaigns Noratlante I, Nestlante I, and Nestlante II not only demonstrated CNEXO's success in creating pathways for industrial-academic collaboration, but they were also fundamental to constructing the submerged territory where the French extractive industry might operate in a near future. The only large oceanographic vessel in France, the *Jean Charcot*, became the common ground where industrial interests in surveying continental margins merged with scientific motivations to study the dynamics of the ocean crust. While industrial goals were no secret, CNEXO managers encouraged the publication of data acquired as a way of explicitly displaying the scientific character of these missions. In France, the oceans' scientific and economic exploration were growing indistinguishable.

CHARTING THE NATIONAL CONFINES
BENEATH THE SEA

No territory is such without being thoroughly charted. Representations are fundamental to visualize its contours, render familiar its geography, assess its economic potential, and control it to the fullest extent. This is as true for the mainland as for the seafloor. Early in 1968, members of CNEXO's Administrative Council decided to prioritize charting France's underwater territories, mobilizing scientific laboratories and industrial institutions into a coordinated and systematic effort.[39] The resulting program not only brought in a new mode of visualizing underwater regions but, most important for ocean sciences, it was crucial to consolidating France's community of marine geoscientists as a proficient, internationally competitive workforce.[40]

Immediately after CNEXO implemented its national ocean policy, it became evident to its stakeholders that France lacked a comprehensive geological, sedimentological, and bathymetrical map of its continental shelf, the basis for identifying likely regions for the accumulation of building materials, minerals, and hydrocarbons. Until then, mapping activities had been undertaken by multiple institutions and laboratories that drew local, nonstandardized charts, leaving vast submerged regions unknown. Those charts did not take into account the existence of natural resources, but instead aimed to study the seafloor's sedimentological composition for geological investigations.

CNEXO's urgency in undertaking a large-scale mapping effort was largely owed to its general director, Yves la Prairie. He considered that the oceanographic agency should not only coordinate ocean sciences but also take on the task of overseeing offshore industrial activities. In his view, CNEXO needed to possess some control over the leases granted to oil, mining, and fishing industries, and he made this known to the prime minister in 1968. By then, the National Assembly was immersed in defining a national organization to grant offshore mining leases to foreign firms, a task compelled by the 1965 Law of the Sea. The government lawyers had almost finished drafting the legal texts when CNEXO was founded. La Prairie interrupted their effort with a persuasive letter to the prime minister, encouraging him to introduce CNEXO into the organization to grant mining leases.[41] In his letter, la Prairie asserted that CNEXO was the institution best suited to determine which regions or types of resources should (or should not) be extracted due to the exhaustive geological knowledge it possessed. Those data would also be crucial to defining the French delegation's position at future UNCLOS meetings (the upcoming one, planned for 1973, would include discussion of the limits of national jurisdiction over ocean space).[42] La Prairie requested that CNEXO take up a key advisory role, with decision-making capabilities, for granting exploration and exploitation leases.[43] The French Law on the Exploration of the Continental Shelf and the Exploitation of Its Natural Resources did not adopt la Prairie's suggestion in detail, but it did assign CNEXO an advisory role. As stated in Article 34, CNEXO had access to the geological, hydrological, and biological data collected during exploration and exploitation activities.[44] From May 1971, the minister of industrial and scientific development reported to la Prairie regarding the oil leases requested (normally from French oil companies), while la Prairie replied by offering his advice on behalf of CNEXO. He always emphasized the need to carefully define a strategy to prevent oil spills and seawater pollution, and he reminded the ministry that all geological, hydrological, meteorological, and biological information had to be provided to CNEXO.[45] In this way, he ensured significant control over any activity happening on the continental shelf and over any resulting data that might possess strategic value.

This governmental advisory role prompted CNEXO's Administrative Council to prioritize, among a long list of ocean projects, the mapping of France's underwater territories.[46] The Atlantic continental shelf was selected as the priority region for its geographic features and economic promise. During previous preliminary surveys, IFP experts had identified thick layers of sediments indicating likely hydrocarbon occurrences. The wide and shallow continental shelf (up to two hundred meters deep, reaching a hundred fifty kilometers wide) facilitated industrial activities like testing exploration technologies, conducting systematic surveys, and eventually installing infrastructures for oil and gas exploitation. Besides hydrocarbon deposits, the region was covered by surface accumulations of sands and gravels, potentially exploitable in the short term for the construction industry.[47] Conversely, the Mediterranean continental shelf initially received less attention because it was narrower (up to sixty kilometers in its widest point) and not as economically promising in the short term. And even though the narrower Mediterranean shelf would require less effort to explore, preliminary surveys had demonstrated that potential hydrocarbon deposits hid at greater depths, around the salt dome region in the Balearic basin. During the first years of CNEXO activities, the Mediterranean's exploration was left mainly in the hands of academic researchers, while industry-academia surveys concentrated on the hexagon's Atlantic coastline.

No efforts were spared for the Atlantic mapping. CNEXO mobilized numerous research teams around the country, transforming the North Atlantic continental shelf's cartography in the first large-scale, coordinated project under French marine geosciences.[48] CNEXO formed the core of a network that connected academic research teams, public institutions, and private companies, and through which all data acquired and jointly interpreted flowed to CNEXO. The program was particularly attractive for those academic geologists willing to align their research lines with the national agenda. CNEXO granted them economic support, facilitated inter-institutional collaborations and—most importantly—enabled them to access industrial-quality data.

At the University of Bordeaux, geologist Michel Vigneaux benefited from CNEXO's cartographical program to foster his research lines. He occupied a prominent advisory role at CNEXO as a member of its Scientific and Technical Committee, where he developed a close friendship with la Prairie. For the national cartographical project, Vigneaux offered his laboratory and research resources to survey the Bay of Biscay, obtaining in exchange seismic profiles of industrial quality from the IFP and 120,000 US dollars from CNEXO, which enabled him to acquire samples of the region's upper sediments using marine geological techniques.[49] Deploying dredges and sediment corers, Vigneaux's team at Bordeaux produced sedimentological and geochemical data that were shared with research groups working with marine geophysical profiles: the geophysicist Pierre Muraour, at the University of Montpellier, and cartographers at the Geological and Mining Research Office (BRGM).

Just as Vigneaux needed geophysical data to organize his campaigns, geophysicists required geological information to precisely interpret the sedimentary formations displayed in their seismic profiles.[50] Similar partnerships were established at other French institutions: at the University of Rennes, geologist Gilbert Boillot and his team worked shoulder to shoulder with the offshore-tech company Geotechnip, with which they exchanged geological data for reflection seismic profiles of the English Channel's eastern region.[51] Geologists at the IFP contributed their *Flexotir* data, acquired during oil prospecting missions, and the vessel *Petit Marie François*, equipped to undertake seismic surveys.[52] Results and data flowed to CNEXO embodied in final reports, which were then passed on to experts at the BRGM, who were responsible for drawing the geological map at 1/1,000,000 scale.

From this joint effort, the contours of the French underwater territory began to take shape. The combination of rock core samples, obtained mostly by academic research teams, and light reflection seismics, which provided information on the seafloor's composition, proved valuable for drawing the features of the seafloor (table 4.1). The 1/1,000,000 geological map, finalized in 1976, showed the different rock formations and structures that made up the continental shelf, offered information about its age and composition, and depicted the region's geomorphic characteristics.

The geological history of the Bay of Biscay, which could now be read through the underwater chart, was as important for understanding the region in the framework of plate tectonics as for assessing its economic prospects. Knowing the evolution of marine sedimentary basins was essential for estimating their hydrocarbon potential: basins recently formed would be poor in hydrocarbons, while an ancient basin, in subsidence for millions of years, offered more positive prospects due to the high weight of accumulated sediments.[53] The convergent economic and scientific interests in the region, intensified by the large quantity of data produced in a short span of time, transformed the Bay of Biscay into the birthplace of French marine geosciences as large-scale, internationally relevant, sciences.

This achievement was symbolized by the international Symposium on the Structural History of the Bay of Biscay, held in December 1970. More than two hundred experts from oil companies, national geological services, and universities from Europe and the United States gathered at

Table 4.1. Contribution of different French institutions to mapping the Atlantic continental shelf

Institution	Contribution (in % of data)	
	Core samples	Seismic data
Universities	2.3	11.3
Bordeaux	16.9	2.3
Caen	8.7	12
Nantes	50.9	22.9
Paris + Rennes	1.1	0
Perpignan	3.8	0
Rouen		
Total universities	83.7	48.5
Other institutions		
IFP	10.7	10.9
BRGM	5.6	24.3
CNEXO	0	16.3
Total other institutions	16.3	51.5

Source: Modified after Boillot, "Des marges continentales Atlantiques."

the IFP's headquarters in Rueil-Malmaison.[54] Organized by le Pichon, from CNEXO, and the IFP's marine geophysicists Lucien Montadert and Jacques Debyser, the symposium manifested how the boundaries between academic and industrial research had blurred. Communications alternated between academic investigations and results obtained during oil exploration. Renowned managers from France's oil industry, like CEPM's President Maurice Leblond, the IFP's General Director André Navarre, and even Yves la Prairie delivered the conference's opening speeches, while petroleum geologists and foreign geoscientists chaired the sessions.[55]

As stated by the organizers, the symposium "demonstrated, if proof is still necessary, that in the field of geology no progress is possible except through close cooperation between fundamental and applied research."[56] French oil companies disclosed information hitherto confidential about the geology and structure of the Atlantic margin, while experts involved in the mapping project had the opportunity to present their results to the international academic community.[57] Le Pichon framed the Bay of Biscay for the first time within the theory of plate tectonics by bringing together the conclusions reached during the mapping project with the results obtained during the CNEXO campaigns Noratlante and Nestlante.[58] In this new framework, the region was conceived as a passive continental margin: a transition area between the continental and oceanic crusts, characterized by thick sedimentary layers and geological inactivity. Active continental margins, on the other hand, corresponded to the boundaries of tectonic plates, where geological activity is high (earthquakes, vulcanism) and are related to subduction processes or convergent plates. The Bay of Biscay, thus, did not correspond to a boundary of tectonic plates but instead appeared after the opening of an ancient rift due to the rotation of the Iberian Peninsula (with the axis in the Pyrenees) during the Mesozoic, about 110 MYA.[59] It was largely a tectonically inactive region, offering optimistic forecasts for industrial exploitation.

From geologists' perspective, the 1970 symposium constituted the founding act of marine geosciences in France, a milestone event in the geological understanding of continental margins and Earth's evolution in the light of plate tectonics.[60] In other words, through their eagerness to explore and exploit natural resources from the seafloor, CNEXO managers not only created the conditions under which geological data and knowledge could flow among national research institutions but also promoted the emergence of physical spaces where this information could be shared and debated with the international community, leading to a fundamental understanding of the history of the oceans.

NAVIGATING THE DIPLOMACY OF THE SEA'S
(UNDELIMITED) BOUNDARIES

Beyond industrial exploitation of mineral resources and fundamental knowledge of the region's tectonics, mapping the Atlantic continental shelf served other national purposes. It became a diplomatic tool, utilized to prevent diplomatic frictions, and a pilot program for undertaking industrial activities over foreign continental shelves.

Shortly after launching the mapping program, la Prairie met an advisor from the British Natural Environment Research Council (NERC), Mr. Fenning, to discuss a cooperative project to explore the central and western English Channel. Like CNEXO, NERC was a research council funded in 1965 to coordinate research and training in environmental sciences. Diplomatic relations between France and the UK were tense regarding the limits of their national territories bordering the English Channel, where both countries held military and economic interests. Negotiations to trace the exact boundaries expanded from 1970 to 1974. The moment was timely to suggest a cooperative undertaking that would prevent misunderstandings about the exploration of the English Channel and about potential political tensions derived from the discovery of natural resources in middle grounds. With the national charting program in place, la Prairie could suggest to Mr. Fenning that France and Britain should harmonize their national mapping programs, thus preventing unnecessary duplication of effort while sharing "the most relevant discoveries" (in natural resources, it was implied). The agreement entailed fostering the organization of joint oceanographic campaigns, establishing a standardized, common geological chart of the region, and collaborating to reach a legal consensus about the seabed's borders.[61] This agreement became the seed of a wider, bilateral collaboration in marine sciences between France and the UK.[62]

In addition to identifying hydrocarbon deposits near the hexagon, CNEXO's mapping program of the Atlantic continental shelf included deploying exploration techniques to identify mineral deposits, like metals or placers. However, from the program's launch, experts at CNEXO's Administrative Council were well aware that, on France's continental shelf, those resources were scarce or absent.[63] Why, then, invest in training experts and developing research methodologies for a resource that would not be economically profitable? As la Prairie plainly expressed, mineral prospecting activities would serve as pilot studies: French experts would be able to develop the skills and technologies needed to exploit underwater regions in overseas and foreign territories.

Seeing the exploration of the oceans as an international race to grab un-discovered marine riches, la Prairie pushed to acquire mining leases over foreign continental shelves even before France had developed the techno-logical capabilities to exploit them.[64] Regions in CNEXO's spotlight were "francophone countries"—that is to say, former colonial territories mostly in central and southern Africa, whose extractive technological capabilities were far behind those of France. CNEXO's Administrative Council did not consider these enterprises as echoes of former power relations, but (in their rhetoric) as the means to bring France's expertise and technologies to less technologically developed nations, especially those with which diplomatic relations ought to be strengthened.[65] They believed that sharing a portion of the economic benefits, such as through acquiring mining leases, or promot-ing the establishment of industries to process raw materials, would enhance bilateral relations. More or less inadvertently, these experts were proposing to replicate a similar industrial schema over the oceans to the one formerly established on mainland colonial territories although, this time, France did not rule over the regions to be exploited. These ambitions began to crystal-lize early in 1971, when CNEXO undertook a global-scale synthesis to define coastal regions favorable to the accumulation of placers like zircon, rutile, ilmenite, gold, or diamonds. In collaboration with the BRGM, the over-arching goal was to facilitate the exploitation of seabed minerals around the world.[66] Exploration over foreign continental shelves began in Senegal, where French academic and industrial experts explored the accumulation of ilmenite, a titanium-iron oxide mineral and valuable source of titanium, with the positive support of the Senegalese government.[67] Yet, before the exploitation of valuable resources was even envisioned, CNEXO changed the direction of its regional strategy—from foreign continental shelves to the high seas. Although France's mining companies and the national geo-logical agency BRGM would continue exploring the economic possibilities of extracting mineral resources in foreign waters, a new, higher priority, was to direct scientific research toward the international high seas, which were emerging as the most promising future sites for human exploration.

THE SEAFLOOR IN DEPTH: KNOWLEDGE, COLLABORATIONS, AND EXPANSION

By 1971, after four years of existence, CNEXO had managed to forge a solid cooperative network between national oil companies, public bodies, and academic geologists by guiding research lines toward shared goals. Com-mercial and economic ambitions drove the exploration of the oceans, while fundamental knowledge of the dynamics of the oceanic crust was a first and central step to charting the existence of valuable resources.

The CNEXO-sponsored oceanographic surveys played with the blurred boundaries of the category "scientific research" as defined in the 1958 Convention on the Continental Shelf. CNEXO organized scientific geophysical surveys to provide valuable data to France's oil industry by pursuing a double agenda: building knowledge about tectonic processes of the oceans' crust while gathering structural and geological data to prepare future oil exploration ventures. This proved to be a canny strategy for exploring the oil potential of continental margins without needing to acquire leases for oil exploration and exploitation, and it generated unprecedented knowledge about Earth's dynamics. Stimulating information exchanges led to situations where geological knowledge of the seafloor was coproduced between oil experts and academic geologists. But beyond their industrial and scientific interests, these missions on the high seas and over foreign territorial waters were also sending a clear message to the international community: France now possessed technological, human, and logistical capabilities to organize large-scale oceanographic campaigns and to engage in cutting-edge research lines. It was a genuine display of how the French government was supporting the growth of ocean sciences embedded in big science settings, and of its ambitions to expand France's presence across the oceans. The examples presented here show France expanding its presence from the French continental shelf to the coastline of Senegal and across the deep North Atlantic. All of those oceanic areas were centers of potential economic interest, where France could demonstrate its international presence through scientific-industrial exploration. CNEXO had become a key player in international ocean diplomacy.

[CHAPTER 5]

Alliances and Hidden Minerals
in the Abyss

It is not down on any map; true places never are.
HERMAN MELVILLE, *MOBY-DICK*

On November 1, 1967, the diplomat Arvid Pardo, Maltese delegate at the United Nations, introduced the imaginary of untapped riches beneath the deep oceans, awaiting discovery. In a famous speech, Pardo presented an alluring description of boundless seafloor treasures opened up via future technological innovation—minerals, sands, and gravels that would cover global consumption for thousands of years, hydrocarbon reserves that promised to sustain the growing global demand for decades. But he combined his picture with a warning against the seafloor's grab and enclosure by the most powerful nations in ocean research. He alerted attendees that some countries were using "their technical competence to achieve near-unbreakable world dominance through predominant control over the sea-bed and the ocean floor."[1] There was a looming risk of perpetuating the colonial world order in the oceans. As Pardo warned, "The process has already started and will lead to a competitive scramble for sovereign right over the land underlying the world's seas and oceans, surpassing in magnitude and in its implication last century's colonial scramble for territory in Asia and Africa."[2]

In Pardo's vision, the seafloor was equivalent to land: a territory to be explored, mapped, colonized, and connected to mainland economies through national claims and technological occupation.[3] From this fear, and drawing inspiration from previous international treaties taking "the benefit of humankind" as a key legal concept, Pardo proposed natural resources on the high seas as "the Common Heritage of [Hu]mankind," that is, belonging to present and future generations, not subject to national appropriation, and reserved exclusively for peaceful purposes.[4] The concept became a powerful landmark throughout the following decade of UN negotiations.

112

It dominated legal discourses on governing global commons, where tensions between sharing and preserving natural resources in the high seas had to be incorporated into an internationally regulated exploitation regime.

Pardo's speech encapsulates the ideas about ocean exploration that, by the end of the sixties, had gained momentum within international politics: the paired concepts of competition and cooperation, global benefit and national profit, sharing and gaining an advantage in commercial exploration. These notions permeated the position that coastal nations adopted on the scientific exploration of the oceans. This chapter focuses on how these ideas and apprehensions shaped France's endeavors to expand toward the deepest regions of the Atlantic and Pacific Oceans. For each of these oceanic scenarios, the chapter analyzes the creation of a CNEXO research base (the Oceanological Center of Brittany and the Pacific Oceanological Center) and a research program (the Franco-American Mid-Ocean Undersea Study, a.k.a. FAMOUS, and the program on polymetallic nodules). These two cooperative projects pushed marine geology into new frontiers by retrieving samples from the ocean floor at thousands of meters' depth, relying on scientific submersibles, prototype mining technologies, and international partners. These cases underscore that, to understand the origin of new capabilities to operate in deep, distant waters (logistical bases, innovative technologies, or new partnerships), it is crucial to consider how oceanographic diplomacy shaped their organization, purpose, and role beyond the sciences. Oceanographic diplomacy involves using scientific research and cooperation in the oceans to promote international relations while advancing a national agenda. In this context, the fears of a postcolonial power struggle in the oceans, the rhetoric of sharing "for a common benefit," or setting the stage for anticipated discoveries, all steered diplomatic discussions and relations.[5]

In his book about American oceanographers during the Cold War, historian Jacob Hamblin analyzed how national security and military priorities shaped international cooperation in ocean sciences. He demonstrated that the US Navy's suspicions toward the Eastern Bloc, coupled with a concern for better understanding the oceanographic capabilities of other nations, led to a peculiar scientific position. Naval officers and oceanographers crafted strategies that balanced sharing and concealment, carefully selecting international partners and projects. In other words, Hamblin showed that we cannot grasp the American participation style in international oceanographic collaboration without understanding the geopolitical and military issues that shaped it.[6] Along similar lines, this chapter contributes to the question Hamblin raised by showing how commercial

interests—intent on gaining new mineral supplies and economic incomes—
also shaped the kind of international relations established to (scientifically)
explore the seafloor.

Going back to the idea that opened this book, assembling the submerged
territory entails scientific effort, technological innovation, and the establishment of national, cooperative networks. For the deep oceans, taming
the underwater territory required forging international alliances, harmonizing objectives, and reaching agreements and commitments that satisfied
all involved parties. The suspicions within the United Nations, the banner
of exploring for a common good, and the competitive eagerness to control potential mineral deposits in offshore waters significantly influenced
France's activities in the open ocean. Through CNEXO, France cautiously
navigated the fears voiced at the United Nations by establishing international agreements, fostering an environment of mutual trust, and sharing
resources, expertise, data, and coastal bases with selected partners such as
the United States or Japan. At the same time, returns materialized in the
form of increased international prestige and information about the deep
seabed's commercial potential.

CNEXO's expansion into the deep floors of the Atlantic and Pacific
Oceans, manifested in the sampling of rocks and polymetallic nodules
thousands of meters deep, presents an excellent case to analyze these dynamics. The chapter begins by detailing the context in which the ideas,
fears, and concepts presented above emerged, then traces how they stimulated CNEXO to play a key role in mediating international relations through
ocean sciences. The Atlantic Ocean's case is tackled through the science
diplomacy relations of France and the US. The creation of the Oceanological Center of Brittany, or COB, resulted from the convergence of both
countries in ocean exploration, embodying paired cooperation and competition at work. The joint mission FAMOUS, with its manned submersibles, consolidated France's presence in the deep Atlantic Ocean under
the aegis of pursuing research for the common benefit of humankind. The
mission enhanced France's standing on the international stage, positioning the country alongside the US as a pioneer in deep ocean technologies.
For the Pacific Ocean, I detail how the French government's interest in assessing the existence of polymetallic nodules led not only to the creation
of an oceanographic base in French Polynesia (the Pacific Oceanological
Center), envisioned as a logistical base for both international and industrial
missions, but also to establishing strategic relations with Japan and Germany so that, together, they could share expeditions, prototype technologies, and data.

INTERNATIONAL FEARS AND ASPIRATIONS IN
FRANCE'S OCEAN DIPLOMACY BLUEPRINT

When steering its future industrial development toward the oceans, France was following a global current that, by the late sixties, was fueled by anxieties about overpopulation, resource erosion, and environmental degradation on land. Global trends showed exponential growth from the start of the century: the world's population doubled, along with the agricultural land cultivated to feed it, while energy consumption more than tripled.[7] The rampant process of decolonization compelled the redefinition of national borders, while hostile international relations between East and West, developed and developing states, jeopardized international political stability. The "tragedy of the commons," a concept biologist Garrett Hardin introduced in 1968, epitomized this feeling, becoming an ominous warning of the catastrophic consequences of a growing population driven by a consumerist, laissez-faire ideology.[8]

At the UN General Assembly, Pardo's advocacy for legislating the still-to-be discovered resources of the high seas as belonging "to the common heritage of mankind" emerged from that context. Arvid Pardo, born in Italy in 1914, graduated in international law and began working at the UN after the end of World War II. He started in the Department of Trusteeship and Non-Self-Governing Territories and worked for the secretariat of the Technical Assistance Board before being appointed Malta's chief diplomat: the first permanent representative of the newly independent country.[9] His 1967 speech addressed Malta's request to bring to the fore, in the UN General Assembly's discussions, the threats posed by new offshore mining technologies, including national appropriation and militarization of the seabed.[10] The Maltese government feared that the most technologically capable nations could secure marine regions beyond national jurisdiction—something that was already happening around its own national waters regarding fishing resources.[11] Illegal fishing was becoming common across the Mediterranean Sea, while oil companies were moving prospecting activities further from the continental shelf. Pardo, voicing Malta's concerns, denounced these initial steps as a sort of oceanic neocolonialism, in which developing nations would fall again under the control of those more technologically developed.[12]

Until that time, the natural resources beyond the continental shelf were *res nullius*; they belonged to the first who took them. In Pardo's speech, the seafloor was openly conceived as a territory to be occupied and governed, the battleground for defining a new world order.[13] Three competing visions, depicted in Pardo's discourse, informed international understanding of this new piece of land: the oceans as "the womb of life," to be safeguarded; the oceans as a new economic frontier, with boundless richness susceptible to national appropriation; and the oceans as the great and still remaining commons of humankind, whose resources should equally benefit all nations—preserving, exploiting, and sharing. Pardo's advocacy for the yet-to-be-discovered resources as belonging to the common heritage of mankind implied a peaceful (and arguably utopian) approach to the oceans' exploitation. As he suggested, an international mechanism defined by the UN might ensure the sharing of economic revenues from offshore mining activities. This source of funding offered a potential solution to world hunger and poverty—and, therefore, directly tackled global anxieties that pervaded the international landscape. Pardo's proposal garnered different reactions. While developing nations sympathized with it, gathering for the first time around a unified position against the excesses of (technologically) developed nations, these later were dazzled by the promises of undiscovered resources hidden in ungoverned regions.[14]

In the 1960s, then, France found itself among those seeking to control the "seabed's boundless treasures," as Pardo had called them. French territories lay scattered around the globe, encompassing deep, yet-unexplored regions beneath the Atlantic, Pacific, and Indian Oceans. Assessing the economic potential of those distant regions required large sets of oceanographic data, collected through logistically complex scientific expeditions. At the same time, the resulting knowledge was key to informing France's position in the legal debates simultaneously taking place. The creation of CNEXO represented a canny opportunity to combine the scientific exploration of the high seas with political priorities. Diplomatic affairs became integrated as one of the institution's main advisory roles, as General Director Yves la Prairie began to build further connections (beyond domestic offshore industries) to transform CNEXO into the main national hub for ocean diplomacy.

At CNEXO's headquarters in Paris, la Prairie established a Department of International Relations and appointed his close colleague and naval commander Alain Sciard its director, and lawyer Georgette Mariani as deputy head.[15] Sciard held a brilliant professional record at the French Navy and had never been involved in administrative matters. Yet la Prairie chose him for the position because of his proficient diplomatic skills—he had demonstrated good relations with academic researchers, military officers, and civil servants, in France and abroad—his mastery of international politics, his deep understanding of marine technologies, and his competence in English. Accepting the offer, Sciard became the mediator between different governmental offices in France and abroad, and advisor for issues related to ocean legislation.[16] For her side, Mariani was a perfect fit as advisor to matters related to the Law of the Sea. Trained in international law at the universities of Harvard, Urbino, and Paris, Mariani was familiar with international diplomacy from her ten-year position as administrator at the Commission of European Community in Brussels.[17] She represented France at UN conventions and committees, reporting to la Prairie about discussions and negotiations that had taken place. He, in turn, reported to Minister of Foreign Affairs Maurice Schumann.[18] Through the International Relations Department, CNEXO became a location where scientific and diplomatic affairs mingled to inform political decisions.[19]

Soon the significance of CNEXO's diplomatic role became evident, as France swiftly needed to define its position in international ocean governance. Pardo's speech had triggered international interest in regulating the global commons, fostering the reopening of negotiations in the UN to legislate the high seas—starting under the ad hoc Seabed Committee.[20] The "common heritage of humankind" became the indisputable principle upon which a mechanism to regulate resource exploitation on the high seas should be built; yet tensions arose when defining the boundaries of international waters. Agreement was reached on the general guidelines of an international regime to govern these regions: only peaceful activities could be undertaken, and the exploration and exploitation of natural resources for the benefit of all humankind was to be developed under an international regime.[21] There would be neither appropriation nor claims of sovereignty over resources or territory, and benefits would be shared among coastal nations.[22] However, the committee failed to specify a fine-grained regulatory mechanism or to draw clear boundaries between national and international waters, fostering the organization of the Third UN Convention on the Law of the Sea (UNCLOS III).[23] The position of coastal nations at UNCLOS III would be defined by their anticipated future ocean technological capabilities, which would enable them to survey, control, and eventually extract natural resources.[24]

At CNEXO, this understanding of the high seas and their resources as a common good stimulated its exploration in cooperative surveys. Unilateral cruises across the high seas, traditionally considered as exercises of national power, were no longer perceived as favorably. At the same time, the (constant) fear of falling behind other nations regarding oceanographic prowess bolstered CNEXO's willingness to share expeditions, cruises, and data. A science diplomacy strategy was taking shape, encapsulated by the metaphor of a race. As Jacques Perrot, vice director and right-hand assistant to la Prairie, affirmed, "Each time more competitors join the race, but the oceanographic adventure is not a 100-meter [race] in which each runner runs in its own corridor; instead it's a long-distance race, where tactics are as important as keeping the breath. We need to take the lead from the very beginning, since once the race has begun, the efforts to leave behind the bulk of racers and find a new place [at the forefront] will be excessive if compared with the efforts required at the beginning."[25]

The race metaphor translated into a quest to establish convenient relations with those nations that, in one way or another, would enable France's access to the deepest ocean floor, with whom it could develop scientific expeditions for the benefit of humankind—that is, avoiding explicit military or commercial implications—and with the means to strengthen France's international power.[26] At CNEXO's Department of International Relations, Sciard maintained close correspondence with scientific attachés at France's embassies abroad, including in the US, Japan, and Germany, who periodically reported to him about foreign oceanographic advancements, missions, and results, annual budgets for oceanography, and even news about activities undertaken by private companies.[27] Reports were passed from Sciard to la Prairie, and from him to the different ministries involved, constituting the basis for discussions of strategies for bilateral and international cooperation.

Other European nations were regarded as delayed in ocean matters, since none had yet established a centralizing structure like CNEXO with which to negotiate.[28] The Inner Six (Belgium, France, Italy, Luxemburg, Netherlands, and West Germany) started meeting in January 1968 to delineate a cooperative program in ocean sciences, but la Prairie criticized the proposed goals as not aligned with CNEXO's ambitions: representatives from universities, who dominated the meetings, pushed for cooperation in "fundamental oceanography" instead of "preparing the future exploitation of the oceans."[29] Conversely the US, the USSR, Japan, the UK, and Canada were deemed potentially conducive partners: the first two were the only nations that had built a centralized ocean policy, while Japan and Canada were investing large budgets in preparation for the oceans' exploitation.[30]

According to CNEXO's approach, a nation's advancement in ocean exploitation was measured not by the economic revenue from a de facto exploitation, but by its investment in ocean *exploration* (table 5.1). The race metaphor installed competition at the level of investing in anticipation of resource extraction—in occupying a leading position "before the race began." In other words, for CNEXO managers, public expenditures in oceanographic research were regarded as the key to estimating the nation's potential to extract marine resources in the near future.

Table 5.1. Public investment in ocean research (in millions of 1970 US dollars)

	1969	1970	1971
US	565.7*	-	332.7
Japan	8	18	27
France (CNEXO)	8.8	12.8	16.2
UK	32.3	-	33.2
West Germany	-	32.3	33
Canada	32.3	28.7	-

Source: Based on the information gathered at CNEXO's Department of International Relations (*Three-fifths funded through military contracts). Data from 1969 and 1970 from La Prairie, "Fiche sur l'effort français en océanologie," January 12, 1970 (ANF, 20160129/325); 1971 data from CNEXO, "Effort financier public poursuivi en matière océanologique par les principales puissances industrielles," September 19, 1972 (ANF, 20160259/318).

Recognizing the US as the world pioneer in ocean exploration, CNEXO first focused on establishing collaborations with American research centers and, along the way, emulating the US in order to achieve its own success. The design of CNEXO's Oceanological Center of Brittany, its new institute for ocean sciences and technology, offers a clear example of this dynamic. French government officials and ocean industry representatives embraced the American model to guide the development of large-scale ocean exploration in France.

CROSSING THE ATLANTIC DIVIDE THROUGH OCEAN SCIENCES

Soon after CNEXO's funding in April 1967, its General Director Yves la Prairie started to reach out to representatives of American ocean sciences.[31] The diplomatic relation between the two countries was rather tense. France had just withdrawn from NATO's integrated military structure, and President de Gaulle undertook during his term in office a constant policy of moving away from the American sphere of influence with the goal of becoming an independent force. Cooperation in ocean exploration could provide a back channel for diplomacy: a means to strengthen diplomatic relations through activities undertaken by nonstate actors.

La Prairie was quick to open a conversation with Edward Wenk Jr., executive secretary of the National Commission on Marine Science, Engineering, and Resources—CNEXO's counterpart in the US.[32] Wenk, who trained as a civil engineer at Johns Hopkins University, had worked as an engineering specialist in submarines for the US Navy before turning to policymaking. In 1959 he was appointed the first science policy advisor to the US Congress, playing diverse advisory roles at the White House in the administrations of Presidents Kennedy and Johnson, including leading the establishment of the National Commission on Marine Science.[33] In their correspondence, Wenk and la Prairie agreed that the concerns and interests of the commission and CNEXO were very similar, which led la Prairie to predict in a private letter to Minister Maurice Schumann, then France's minister of scientific research and atomic and space affairs, likely future cooperation with the US.[34]

In March 1968, la Prairie traveled to the US on an official trip to participate in the second meeting of UN Experts in Marine Science and Technology, held in New York. He used his trip to meet Wenk and prominent experts in American oceanography—both scientists and policymakers. Wenk organized la Prairie's visit to the White House, where he met the American Vice President Hubert H. Humphrey, overseer of American ocean policy under Johnson's administration.[35] Despite diplomatic tensions between the two countries, Humphrey warmly welcomed la Prairie, exclaiming that he was and would always be "deeply Francophile."[36] They discussed how their ambitions in ocean exploration aligned, envisioning Franco-American cooperation.

In addition to diplomatic efforts, la Prairie's trip had another purpose. CNEXO was planning to establish France's largest marine science and technology research institute on its Atlantic shore. Until then, in France, marine research had developed at universities or in coastal stations attached to them, such as in the laboratory of marine geology at Villefranche-sur-Mer and the Universities of Brest, Rennes, and Montpellier. Research disciplines like marine biology, marine geology, or physical oceanography were segregated into different centers or departments, which made it difficult for their researchers to establish multidisciplinary cooperation. CNEXO officials believed that having a large and advanced research center, gathering diverse disciplines and research approaches, was crucial for coordinating national research and advancing ocean exploration. Thus, in his tour overseas, la Prairie also aimed to learn from American research centers by visiting their facilities, talking to scientists, and observing their activities, to bring back insights for success. La Prairie openly expressed his interest, so Wenk organized private tours for him across the East and West Coasts. They visited Lamont, the renowned center in marine geosciences located in New York state; Woods Hole Oceanographic Institution, a major hub for marine biology, oceanography, and marine geology on the shore of Cape Cod; and the Dorothy and Lewis Rosenstiel School of Marine and Atmospheric Science, nestled in the shores of Miami, Florida, and well known for its multidisciplinary research into tropical marine environments. On the West Coast, la Prairie visited the Scripps Institution of Oceanography, the largest center for all kinds of ocean research facing the Pacific Ocean, off southern California. The institute-hopping tour served to foster an atmosphere of mutual trust, building solid relations between la Prairie and the directors of American research centers. Upon returning to CNEXO's Parisian headquarters, la Prairie sent them letters expressing gratitude for their hospitality and valuable insights about the centers' management, organization, and technologies. In a letter to Lamont's associate director, J. Lamar Worzel, la Prairie even emphasized his hopes to build a research center as successful as theirs.[37]

As the representative of French ocean sciences abroad, la Prairie was showcasing France's high regard for the American way of organizing and conducting marine research—a friendly gesture that could facilitate cooperation. Yet his proclaimed admiration was not just a matter of words. CNEXO adopted the organization of American research centers to build its own Centre Océanologique de Bretagne (COB), the Oceanological Center of Brittany.[38] In so doing, France (through CNEXO) was assuming that the American research model was the optimal approach to streamline ocean exploration. This model consisted of large research centers that required substantial funding, where various stakeholders (like the military, offshore industries, and government offices) converged to support research teams and programs. These institutions managed and maintained the American oceanographic fleet and research technologies, with universities relying on them for organizing cruises. They served as launching points for large-scale expeditions, through which researchers gathered extensive datasets. Upon return, data were processed, exchanged, published, and stored. This organization and research style was different from the one developed at laboratories and departments in France before CNEXO's foundation—the one that CNEXO officials were determined to change.

Inaugurated in 1971, the COB sat in the coastal city of Brest, on the western tip of Brittany, overlooking the Atlantic Ocean. The location was ideal: besides facilitating the cooperative exploration of resources in the national continental shelf, it would foster international cooperation on the high seas.[39] The new center was designed as a major eastern Atlantic hub for international oceanography, serving as a port of departure, arrival, and stopover for oceanographic cruises. As stated in the journal *Science* after the center's inauguration, the COB was "both a symbol and the first fruits of a new concentrated national attack on oceanographic problems." The article concluded that "[France] seems to feel that in oceanography an investment of this size will keep [it] in a competitive position about other nations."[40]

The COB's design, from the general distribution of buildings to the organization of scientific research, owed much to Franco-American diplomatic relations. La Prairie had succeeded in bringing home insights about the organization of American research centers. To this feedback he added the reports that the scientific attaché at France's US Embassy periodically sent him about American budgets, resources, and policies addressed to exploring the oceans. As a result, CNEXO had access to large amounts of information about American research institutions: bird's-eye-view maps of the Lamont campus and Woods Hole, detailing the activities developed in each building; a summary of the space that Scripps devoted to diverse activities (administration, research, and support); and a breakdown of the number of rooms per building at Scripps.[41] The COB complex's layout was inspired by all of them. After finding the main building, which hosted administrative activities and a conference room, the visitor would find the new National Office of Oceanic Data, located in a central position, so it could be easily accessed from the research buildings that surrounded it. The eastern corner hosted the "industry-type" halls, and closer to the ocean sat the reception and catering building, enjoying spectacular views of the bay.[42] To gauge the COB's international stance, CNEXO officials compared the allocation of space for various activities at the new center with that at the Canadian Bedford Institute of Oceanography and at the American Scripps. While Scripps allocated three times more area to research labs than the COB, the latter dedicated more space than the Bedford (7,710 m² versus 5,300 m²). The COB nearly matched Scripps in industry halls (9,075 m² versus 10,500 m²) and slightly exceeded both North American centers in area allocated for logistical activities.[43] Like the American research centers, the COB hosted the oceanographic fleet designed to operate across the Atlantic Ocean. This included the R/V *Jean Charcot* and the R/V *Noroit*, both longer than twenty-five meters and capable of operating up to four months at sea, and the brand-new manned submersible *Cyana*, capable of diving to 3,000 meters deep.[44] At the COB, CNEXO started to emulate the American model of marine science: acquiring massive quantities of data during long, far-reaching campaigns that mobilized a large human force and a complex logistical setting.[45] To facilitate communication between scientific disciplines, the French research program was divided into the Division of Solid Environment, Division of Fluid Environment, and Division of Living Environment. At the COB, France's oceanographic field expanded from local, national regions across the high seas via the North Atlantic Ocean.

In parallel to the COB's construction and inauguration, relationships between CNEXO and its American counterparts intensified. The talks opened in 1968 between la Prairie and the American Vice President Humphrey produced a bilateral agreement to explore the oceans in February 1970.[46] The agreement's principles reflected the interest of both nations to explore the great depths, framed in the discussions that were simultaneously taking place at the United Nations. The first article stated that any cooperative project should, above all, "advance study and effective utilization of the sea for the benefit of all men."[47] This exhibition of good will was phrased in a rhetoric inspired by the "common heritage of humankind" concept; yet the Franco-American agreement also reflected the position adopted by both countries at the forefront of ocean exploration, (unilaterally) recognizing themselves as its leaders.[48] Because its spokesmen (la Prairie at CNEXO, Wenk and Humphrey at NOAA) viewed their institutions as equivalents, an equitable partnership in carefully selected projects could bring mutual benefits in exploring new territories and in exchanging technologies, expertise, and scientific knowledge.[49] Therefore, with the agreement's signature, CNEXO continued building an environment of mutual trust with the Americans, in which both communities could exchange information and design joint programs to access the great depths side by side. From this framework emerged the idea of organizing a cooperative project deploying the most advanced deep-sea research technology: manned submersibles.

SHARING SUBMERSIBLES FOR THE DEEPEST GEOLOGY

Marine geology and manned submersibles converged for the first time in the 1972 Franco-American Mid-Ocean Undersea Study (FAMOUS), marking the inaugural exploration of plate tectonics underwater. During the sixties the topic had gained significant traction in the Earth sciences community, following publications that proved seafloor spreading, continental drift, and plate tectonics in a unifying, global theory.[50] Rifts, deep-sea regions where new ocean crust emerges, were in the spotlight of engineers and geophysicists around the world, who competed to develop ingenious technologies to study these areas in situ. Depth was the challenging factor, as it was almost impossible to conduct studies from ships or to use geophysical techniques with sufficiently high resolution. It was necessary to go down and see, to recover samples and to take pictures—something that could only be achieved by using manned submersibles capable of diving thousands of meters deep.[51]

By 1970, only seven manned submersibles could operate at depths greater than 1,830 meters, and almost all of them belonged to the military-industrial complex. The US Navy owned the *Trieste II*, the *Turtle*, and the *Sea Cliff*; while Lockheed's *Deep Quest* was available for military and commercial operations in the US. The *Alvin*, originally designed for military missions at greater depths, was now operated primarily for marine biological research at Woods Hole (table 5.2).[52]

France, though only beginning to make its mark in ocean exploration, possessed two manned submersibles for great depths: the navy operated the bathyscaphe *Archimède*, the oldest of all yet extensively used; while Jacques Cousteau's team had designed the small *Cyana*. Commissioned in 1961, the *Archimède* was not initially intended for scientific missions but rather to set a world record by reaching the oceans' deepest regions (one more example of the international oceanic race, akin to the fervor to explore outer space). At the start of the seventies, the *Archimède* had completed more than a hundred dives around the world's oceans and held three world records for depth. Cousteau's team designed the *Cyana* to explore deep oceanic regions; CNEXO and the DGRST had supported its construction economically and, after its launching, it transitioned under CNEXO's management at the COB.[53]

Xavier le Pichon, marine geoscientist at CNEXO, had harbored the dream of studying the formation of new oceanic crust in situ for more than ten years. Now, CNEXO's ambitious international oceanographic policy offered him a perfect framework to develop such a purposeful research program.[54] After informal discussions with his American colleagues, le Pichon suggested to Director la Prairie a bilateral project to study the Atlantic mid-ocean ridge with the Americans. The topic was compelling due to the unknown nature of the region: Did ocean rifts look like narrow, deep gorges? Or like wide, soft valleys? Could they sustain any living organisms?

Table 5.2. Manned submersibles in 1976 with an operating depth limit exceeding 1.83 kilometers

Submersible	Country	Operator	Year of launch	Operating depth limit (m)
Trieste II	USA	US Navy	1964	6,100
Alvin	USA	WHOI	1964	3,660
Deep Quest	USA	Lockheed Missiles & Space Company	1967	2,440
Sea Cliff	USA	US Navy	1968	1,980
Turtle	USA	US Navy	1968	1,980
Archimède	France	French Navy	1961	9,300
Cyana	France	CNEXO	1970	3,000

Source: Modified after Heirtzler and Grassle, "Deep-Sea Research by Manned Submersibles."

La Prairie grasped quickly the importance of a Franco-American coopera-tive program designed to dive into the depths of these unexplored regions. Operating alongside the world leader could give France implicit stature at the forefront of ocean exploration, while also strengthening diplomatic re-lations between the two nations. Sharing manned submersibles—advanced research technologies, as useful for scientific exploration as for military and industrial enterprises—was an excellent way to inaugurate their bi-lateral agreement of cooperation: it would symbolize their mutual trust, while at the same time making history by achieving a milestone in seafloor exploration.

The FAMOUS expedition, designed to study phenomena occurring at the edge of tectonic plates in the mid-Atlantic ridge, resulted from these ambitions.[55] The program's focus, however, soon shifted from innovative science to technological prowess: the use of manned submersibles became the program's masterpiece, indicating that priority was to direct interna-tional attention to their combined oceanographic capabilities. In the words of Alain Sciard, CNEXO leader on international relations, FAMOUS was intended to "strike the opinion of the scientific community and of the gen-eral public, which can only be profitable for us."[56] To that end, National Geographic ensured massive media coverage from the program's inception to its end.

France contributed the bathyscaphe *Archimède* and the submersible *Cyana* to the project, while the US brought the *Alvin*. All three featured a main chamber which maintained atmospheric temperature and pressure, so the occupants did not need to use special suits or undergo decompres-sion procedures. All three were also equipped with a mechanical arm and clamp capable of ripping off rock samples as well as underwater cameras and sonars. The military bathyscaphe *Archimède* had been designed to go beyond 9,000 meters of depth. Twenty-two meters long and eight meters in height, it was the biggest bathyscaphe used in the FAMOUS mission. It could accommodate three men onboard. The brand-new *Cyana* constituted the evolution of light bathyscaphes that Cousteau and his team had been designing for more than a decade. Even though smaller than the *Archimède* (it was one-fourth its length and one-third its height), it could also accom-modate three experts: the pilot, a copilot, and a scientist, who could spend up to ten hours diving underwater. At CNEXO, le Pichon led the scientific team while Jean-Claude Riffaud, former naval officer and CNEXO's man-ager of fleet operations, organized the logistics. Their American counterpart was James Heirtzler, oceanographer at Woods Hole and le Pichon's mentor during his earlier training in the US. Together, they defined the mission's goals and schedule, identified the sites to be surveyed, and planned the dives. FAMOUS stimulated the exchange of researchers between France and the US and, with them, the transfer of expertise, technologies, and scientific knowledge. The team of five French geoscientists toured American research centers and visited the facilities housing the *Alvin* numerous times.[57] Beyond learning from the American institutions, le Pichon and his colleagues benefited from exclusive training opportunities; they conducted fieldwork in Iceland in order to become familiar with a landscape similar to the environment they would confront below three thousand meters of seawater, and they dove onboard the *Archimède* in the Mediterranean to simulate the diving conditions they would be exposed to in the deep Atlantic.

On August 2, 1973, the first phase of project FAMOUS began.[58] The *Archimède*, with le Pichon onboard, dove more than 2,700 meters deep at the mid-Atlantic ridge, capturing for the first time images of oceanic crust in formation. There was no precedent for such images. After the success of the first dive, the Franco-American team organized the second stage of the program, starting the following summer, when the three submersibles were simultaneously mobilized to explore different regions: the *Cyana* explored the rough territory of the transforming fault, while the *Archimède* studied the transition zone between the fault and the rift, and the *Alvin* dove into the rift itself (figs. 13 and 14). Each submersible completed more than a dozen dives, resulting in 167 sampling operations. Over the course of 200 hours beneath the ocean, the astonished geologists brought two tons of volcanic rock and tens of thousands of underwater photographs to the surface. During the following years, American and French experts kept on meeting to analyze, interpret, and publish the results. FAMOUS provided proof that oceanic crust forms through volcanic activity in the ocean floor, along mid-ocean ridges; at the same time, it strengthened collaborations in marine geology between the two countries.

Studying plate tectonics through the synergic use of three submersibles—the *Alvin, Cyana*, and *Archimède*—constituted a complete breakthrough in marine geological research.[59] This approach allowed geologists to explore the depths of the ocean in an unprecedented way: now they were able to see and experience the seabed firsthand, traversing its mountains, plains, and faults and using mechanical clamps to collect samples of sediment, living organisms, and rocks. The precise positioning of the submersibles also allowed for highly accurate charting, giving researchers a greater understanding of the mysteries of the deep. FAMOUS opened up a whole new way of studying the oceanic seafloor, akin to mining exploration on land, and paved the way for future marine research endeavors.

Le Pichon spotted in CNEXO's support an opportunity to replicate this type of bilateral collaboration with other countries. He believed he could

Figure 13. The deep-sea diving submersible *Cyana* hanging from the stern of an oceanographic vessel during the FAMOUS project. Designed in 1966 by Jacques Cousteau to reach great depths, this "diving saucer" remained in operation until 2003 under CNEXO and its successor, IFREMER. Photo courtesy of Woods Hole Oceanographic Institution Archives, © Woods Hole Oceanographic Institution. Reprinted with permission of Woods Hole Oceanographic Institution.

persuade CNEXO officers again to continue studying tectonic processes from underwater. Given its location on the Pacific Ring of Fire, a highly active tectonic region, Japan was in his spotlight. Following le Pichon's initiative, a decade later, in 1984, the French-Japanese KAIKO expedition was launched. KAIKO was equivalent to FAMOUS in aims and methods. Using the brand-new French manned submersible *Nautile*, it studied geodynamic processes in deep trenches around Japan—areas that, for le Pichon, were geologically unique.[60] This time, instead of rifting processes in which the ocean crust splits apart, the international team studied a convergent margin, where one tectonic plate dives beneath another. If FAMOUS demonstrated that France could stand with the US at the forefront of ocean exploration, KAIKO marked the inauguration of large-scale marine geosciences in

Figure 14. The bathyscaphe *Archimède* on the surface, with people working on deck during project FAMOUS. Originally owned by the French Navy, it was transferred to CNEXO in 1969 to manage its scientific operations. The FAMOUS project marked the end of the submersible's career, which included more than 200 dives worldwide between 1961 and 1974. Photo courtesy of Woods Hole Oceanographic Institution Archives, © Woods Hole Oceanographic Institution. Reprinted with permission of Woods Hole Oceanographic Institution.

Japan. For marine geologist Asahiko Taira, then at the Ocean Research Institute of the University of Tokyo, KAIKO introduced a nascent community of marine geoscientists into the latest research technologies for mapping, sounding, and diving to the seafloor.[61] It provided a new understanding of the deep geology and life forms that surrounded Japan, and it stimulated the country's appetite for integrating itself into the international current of large-scale seafloor exploration. Ten years after FAMOUS, France seemed to have achieved its aim of becoming a global leader in ocean exploration, free from reliance on US guidance. In KAIKO, CNEXO—rebaptized that same year as the French Research Institute for Exploitation of the Sea, IFREMER—adopted the role of catalyst to introduce research methods into other scientific communities. Yet the Franco-Japanese program also demonstrated that cooperation on equal footing extended beyond Western oceanographic powers, in the distant Pacific Ocean.

The main objectives of the French-American FAMOUS, as we saw, were certainly scientific. Some of its results, however, reflected economic motivations. FAMOUS was designed with future industrial applications in deep-sea mining and other exploitative ventures in mind, since rifting areas could create cracks prone to the accumulation of valuable minerals (mining companies had previously identified similar cracks in the Red Sea).[62] Hydrothermal environments, like the areas explored by the three submersibles, were considered rich in mineral resources: The French diving team reported the existence of dark crusts of manganese oxides in the top and southern face of the hydrothermal hills, travertines in the northern face, and red, yellow, and greenish minerals composed by oxides, iron, and carbonates near the mouths of the hydrothermal chimneys.[63]

In the framework of international discussions taking place at the UN on legislating the high seas and their potential resources, FAMOUS can be read through the lens of a Franco-American diplomatic strategy. The choice of international waters as playgrounds for scientific cooperation and the avoidance of explicit commercial goals reflected a joint geopolitical decision.[64] By moving away from national waters, the French and American delegations avoided any kind of friction related to the economic exploitation of regions under national jurisdiction, attributing to their joint ventures the prestigious label of supporting basic research for a common benefit or heritage. But France was not oblivious to American industrial activities. Far from it, CNEXO was following with interest the new ventures of the American hard mining industry to test mining technologies at great depths. The quest to efficiently exploit polymetallic nodules in the high seas was among the main economic interests of the US—and therefore also a significant matter for France.

TAMING DISTANT OCEANS: THE QUEST
FOR ABYSSAL MINERALS

By the mid-sixties, polymetallic nodules—spherical mineral concretions formed by concentric layers of manganese and iron hydroxides—came to be perceived as the most economically valuable sediment of the deep ocean floor.[65] France, through CNEXO, set major expectations in this natural resource that led to new international partnerships, joint industrial-scientific organizations, and scientific understanding of the seabed. Yet the quest for deep-sea nodules became a story about anticipating new and imminent economic ventures that, in the end, never arrived.

In a period of rampant techno-optimism about the promises of the ocean, the mining industry spotted these spheres as the potential supply of the world's needs in metals for a million years.[66] This perception spread outward from Ambassador Pardo's 1967 speech to the UN General Assembly, where he offered what proved to be a quixotic promise: that nodules on the Pacific Ocean floor contained reserves of manganese and cobalt enough to cover two hundred thousand years of consumption, nickel for a hundred fifty thousand years, and aluminum for twenty thousand years, along with reserves of copper, zirconium, iron, titanium, magnesium, lead, vanadium . . . that could also serve future generations.[67] Pardo's estimates relied on the 1965 book by American mining engineer John L. Mero, *The Mineral Resources of the Sea*, which had become a reference for the mining industry.[68] That polymetallic nodules existed on the Pacific Ocean floor was not new, as they had been discovered in 1872 during the *Challenger*'s scientific expedition.[69] Yet Mero was the first who affirmed that those concretions were potential major mines of manganese, nickel, and copper. He predicted that nodules, with concentrations of up to 50 kilograms per square meter, covered about 20 percent of the deep Pacific Ocean floor, at depths between 1,500 and 6,000 meters.[70] More alluring were his economic estimates. Mero affirmed that, via technological innovation, by 1980 polymetallic nodules could be mined, transported to port, and processed for about 28.5 US dollars per ton, whereas their gross commercial value would range from 40 to 100 US dollars per ton.[71]

Mero's projections created a stir among mining industries, since his work offered evidence that exploiting the deep ocean floor at industrial scale would be feasible in the short term. At CNEXO, the Administrative Council started following with interest the mining initiatives of foreign companies (as well as the simultaneous discussions taking place at the UN Seabed Committee).[72] In 1968, the American company Tenneco openly expressed its intention to exploit nodules by creating a specialized firm, Deepsea Venture Inc. Two years later, the American company Glomar Marine Inc., specialized in developing offshore-oil drilling technologies, joined the exploration of the deep Pacific, while the German firm Metal Gesellschaft AG associated in consortium with Deepsea Venture Inc. to start the first prospecting survey across the North Pacific. On the opposite shore, the Japanese government launched a three-year program to explore mineral resources, from its territorial waters down to 4,000 meters deep in the South Pacific, going through the Izu and Bonin Archipelagoes.[73]

At CNEXO, Yves la Prairie understood that polymetallic nodules could be economically relevant. However, he held serious suspicions about whether these American ventures mainly constituted an excuse by the US government to stake a strategically advantageous claim in the Pacific before the new international legislation was enacted, regardless of the nodules' worth.[74] Yet, whether for the economic or strategic value of this resource, France's overseas territories in the South Pacific (French Polynesia, the islands of Wallis and Futuna, and New Caledonia) offered an opportunity to launch equivalent surveys across the deep oceans. Georgette Mariani, representative on the UN Seabed Committee, and Alain Sciard, head of CNEXO's international relations, issued an internal report arguing for the need to devote more effort to exploring the economic potential of these regions, which in the near future could fall under international jurisdiction. Although exploitation was not envisioned in the short term, it was essential to demonstrate as soon as possible to those nations that had already launched studies or developed exploitation technologies that France was at the forefront "when the first fruits could be picked up," as they phrased it.[75] In this framework, CNEXO managers decided to approach suitable foreign counterparts to establish relationships to explore deep regions potentially rich in nodules, to test mining technologies, and to learn from their expertise. Japan appeared an attractive partner because, in addition to its location at the Pacific's core, it was becoming heavily engaged in exploring the mineral possibilities of the ocean floor.

In April 1970, la Prairie visited Japan during a first official trip, accompanied by a French delegation including his most trusted colleagues.[76] The mission's goals were to establish contact with renowned experts in Japanese oceanography, study the structure of the Japanese oceanographic policy (by comparing their programs and goals), visit scientific and industrial facilities, and lay the groundwork for future cooperation. Internationally, Japan stood out for its proficiency in fishing and aquaculture, yet the CNEXO delegation was equally interested in its programs of mineral and hydrocarbon exploitation. Japan had traditionally extracted surface minerals from shallow waters, like coal and magnetite-rich sands, locally mined since the mid-nineteenth century. But now, the country was rapidly moving toward large mining operations in the high seas, allocating significant public and private investments to extract placers and minerals in the near future. The Geological Survey of Japan had already charted about 200,000 square kilometers of its continental shelf (out of 280,000 km²), while private firms such as Sumitomo had undertaken the exploration of nodule deposits thousands of meters deep in the central and western Pacific.

The French party was toured around the Fishing University of Tokyo and the laboratories devoted to marine biology at the University of Tokyo, the offshore oil exploitation company Kubiki, and the headquarters of the building company Taisei Construction. From Mitsubishi Heavy Industries, the party learned about the company's activities in the offshore, while at Sumitomo, the most important mining company in Japan, Japanese experts presented their program to extract deep-sea nodules.

Japanese engineers bid the French team farewell by offering them some nodules as a souvenir.[77]

This last visit, coupled with the internal reports Sumitomo provided to la Prairie, made the French delegation aware that the nodule-mining industry was a serious bet for the near future. As la Prairie affirmed in an internal report, the trip to Japan had opened his eyes: He was now convinced that the American enterprises undertaken by, for instance, Deepsea Venture Inc., were not a geopolitical pretext but a serious investment for a new source of income. Japanese studies showed that polymetallic nodules could cover industrial needs in nickel, cobalt, and manganese for two decades. If shredded, nodules could be used in filters to prevent atmospheric pollution by sulfurous gases, and first estimates demonstrated that the costs of exploitation could be easily covered within the first few years of extraction.[78]

For CNEXO, Japan became a desirable partner for formalized bilateral collaborations. In his report after the trip, la Prairie admitted his surprise in discovering how aligned the priorities of both countries were. Japanese experts were familiar with CNEXO's "Blue Book"—CNEXO's scientific agenda—which had been translated into Japanese and could be found across their oceanographic institutions. The overall impression, according to la Prairie, was that "among all countries, Japan is the most conscious of the importance of marine resources to its economy, and the most committed to exploiting them as soon as possible."[79]

Franco-Japanese relations in ocean sciences intensified after that trip. As a token of confidence, the Japanese invited CNEXO engineer Michel Gauthier onboard the vessel *Chyoda Maru II* for a mission organized by the Japan Resources Association.[80] The mission's goal was to assess the effectiveness of a new dredging procedure to exploit nodules, the "Continuous Line Bucket System" or CBL. It consisted of a long loop of cable to which specially designed dredge buckets were attached at intervals of twenty-five to fifty meters. The mining ship pulled the cable, so that the buckets could descend and be dragged along the seabed to scoop up nodules. The mission offered CNEXO the opportunity to observe and learn from the Japanese mining techniques as well as from the setup of their vessels and laboratories.[81]

Predicting successful future exchanges in ocean sciences, France and Japan signed an agreement of bilateral collaboration in marine research on July 2, 1974, based on the broader Agreement of Scientific and Technological Cooperation between the Japanese and the French governments. Both countries agreed to exchange information relevant to the exploitation of the oceans, with a special focus on polymetallic nodules and fisheries.[82] Diving technologies, coastal management, marine infrastructures, and instruments for underwater observation were also on the list. Once this relationship was secured, la Prairie expressed his thoughts out loud, speculating that "the conquest of the Pacific" would be headed by three nations: the US with the Scripps Institution of Oceanography on the California coastline, Japan with its planned oceanographic center on the island of Okinawa, and France with CNEXO.[83] But, to achieve it, CNEXO needed a stable position in the Pacific—thus, its managers began to pull strings to create an oceanographic base in French Polynesia.

FRANCE'S SCIENTIFIC GATEWAY TO
THE PACIFIC OCEAN

Two years before CNEXO's founding, in January 1965, Minister of Research Gaston Palewski had already invoked the idea of establishing an oceanographic center in French Polynesia. Soon after, during an official visit to its capital city, Papeete, on the island of Tahiti, President Charles de Gaulle proudly announced this plan, appealing to the maritime vocation of Polynesian people—a cliché that the French government frequently used to engage public opinion on its ocean policy.[84] Political reasons to establish an oceanographic base precisely there extended beyond industrial motives—because of French nuclear tests.

France started conducting nuclear tests in February 1960, detonating atomic bombs in the Algerian Sahara. In 1964, after Algeria's independence and the need to find suitable grounds to conduct even bigger explosions (a hundred times stronger than the Hiroshima atomic bomb), French authorities moved their testing grounds to the Pacific Ocean. In this move, they followed the lead of the US and the UK, which had established their test bases in the Marshall and Kiribati Archipelagoes. In 1964, the French government founded the Pacific Experimental Center in Papeete, which became the core of a massive infrastructure across multiple archipelagoes and atolls, expanding up to 1,600 kilometers from Tahiti.[85] More than ninety thousand experts moved to French Polynesia to work on the nuclear test apparatus: military troops, meteorologists, oceanographers, divers, engineers, electricians, workers, and administrative personnel, increasing the region's population by 20 percent. The industry around nuclear testing triggered an economic boom. Ports were refurbished and airstrips built, routes between the metropole and the Pacific grew more frequent, and new economies emerged to feed and house the growing population. By the onset of the seventies the GDP had tripled, and the region's economy was sustained by the Pacific Experimental Center.[86] Nuclear tests began in 1966 and lasted for thirty years. Initially, detonations were atmospheric: forty-one nuclear bombs exploded in the open air, whipping up atolls, destroying coral reefs, and exposing the local populations to high levels of radiation.[87] In 1974, tests moved underground, where explosions happened inside wells a hundred meters below the surface—leading to the subsidence of the Moruroa Atoll, the destruction of surface lagoons, and the pollution of seawater due to accidental leaks.

The French government grew concerned that, if nuclear tests came to an end, the region would suffer a brutal regression in its lifestyle. French Polynesia had two options for its future: either turning to tourism or investing in mining underwater minerals. Foreign companies were already involved in both domains, and French authorities feared losing such a strategic position in oceanographic terms.[88] The UN Seabed Committee had recently suggested a legal regime in which coastal states were responsible for the trusteeship of their surrounding waters and for evaluating exploitation projects presented by foreign nations and companies.[89] Establishing a stable oceanographic base in Tahiti would ensure keeping control of exploration and production permits granted to foreign bodies.

In June 1971, the French government published its decision to establish an oceanographic center in Tahiti: the Pacific Oceanological Center (*Centre Océanologique du Pacifique*). Under CNEXO's management, the institution would focus on studies with industrial applications in fisheries, aquaculture, and exploitation of mineral resources. It was openly envisioned as both a logistical base to explore and assess the abundance of nodule deposits in the South Pacific and as a major actor in CNEXO's science diplomacy agenda.[90] Because of Tahiti's isolation in the middle of the Pacific, CNEXO expected that numerous foreign research vessels would make their supply stops there. This would enable the French to keep track of foreign progress in ocean sciences and technologies and would facilitate scientific and technical exchanges.[91] Although the center took eight years of work to become fully operational due to budget shortages, from the laying of its foundation CNEXO gained an open door to the Pacific and a mooring spot for its high-seas expeditions.

Before the construction of the Polynesian center was made public, CNEXO had been discreet in announcing its campaigns to probe the region's potential in nodule deposits. Instead of organizing its own missions, marine geophysicists from CNEXO had participated in French naval activities across the South Pacific addressed at recovering polymetallic nodules. In December 1970, for instance, la Prairie entrusted geologist Jean-Marie Auzende with the direction of the mission Tahino I, onboard the French Navy vessel *La Coquille*. Auzende traveled in a military aircraft to Papeete and, once onboard the vessel, was responsible for supervising technical operations: dredging nodules and taking underwater images of the deposits near the islands of Tuamotu.[92] The ship recovered more than five tons of nodules, which were sent for analysis to the French mining firm Le Nickel. Sediments around the French overseas territories were promising: they contained much more cobalt than the nodules described by John Mero in his 1965 book (although they were poorer in nickel and copper). Now CNEXO's priority was to identify manganese reserves across the South Pacific for their potential strategic interest. Neither the US nor Europe possessed manganese reserves, yet manganese was a key asset for the steel industry. In 1968 alone, France had spent 21.5 million US dollars importing 890,000 tons of this mineral. Identifying nodule deposits rich in manganese, therefore, was a strategic priority.[93]

In 1971, CNEXO launched an ambitious industrial program to exhaustively explore potential nodule deposits in the South Pacific, first around Tahiti's coastal waters, expanding later to the high seas. This was pursued in collaboration with experts from the company Le Nickel and the French Atomic Energy Commission (CEA), which joined CNEXO in March 1972 to implement deep-sea mining technologies.[94] Thus France's nodule consortium crystallized in 1974 as the French Association for the Study and Research on Oceanic Nodules (AFERNOD), constituted by Le Nickel, the CEA, and the shipyards of France-Dunkerque. The group would be responsible of prospecting, exploration, technological development, international negotiations, and relations with fundamental research related to polymetallic nodules.[95]

Unlike private consortia in foreign countries, AFERNOD was a mixed public and private organization, responsible for centralizing both the scientific and industrial work carried out and the results obtained at a national level. Between 1970 and 1992, the French government contributed 95 percent of AFERNOD's budget—that is, more than 130 million US dollars—in its quest for economically profitable deposits of polymetallic nodules, either by direct investments or through the budget of the public organizations CEA and CNEXO. This demonstrates once again the particular strategy behind France's ocean policy: strengthening relations between academic research and industrial production, and creating networks that brought together private and public firms under a state-led structure (like that achieved with the offshore oil industry).

The team soon launched campaigns to survey the abyssal plain of the South Pacific and to test its own undersea mining technologies. Its new "sampler" was a free-diving device composed of two mechanical jaws that descended to the seabed by gravity. In contact with the bottom, the jaws were released to close, recovering samples from sediment and—if lucky—nodules. The machine could return autonomously to the surface using floating glass spheres, to be recovered by the ship using flags, lights, or sonar beacons. During 1972, CNEXO geoscientists released more than five hundred samplers with almost no losses. However, the Japanese mining system CBL seemed to be more efficient. AFERNOD leaders soon realized that only modest progress was possible without international cooperation, which led them to resort to an industrial consortium constituted by Japanese, American, Canadian, Australian, and German institutions. The group pooled its efforts, research resources, and technologies to study promising mining grounds in international ocean waters.

Back in France, the team of geoscientists at the brand-new COB were not satisfied with this industrial enthusiasm. According to Xavier le Pichon, the scientific priority was to keep on exploring continental margins and incorporating knowledge acquired into the new framework of plate tectonics. CNEXO had been designed to advance fundamental knowledge on the oceans that could later serve economic interests, so, he thought, the exploration of the Pacific Ocean floor should be definitely oriented toward its geological understanding, more so than testing mining systems. Fearing the loss of support from CNEXO's scientific force, la Prairie agreed to use industrial campaigns across the Pacific to obtain geological data, thus providing geoscientists unprecedented access to systematic large-scale campaigns across the high seas.[96] Beginning that October, the series of missions called Transpac traversed the South Pacific from side to side—from Panama to Tahiti, from Moruroa to Lima.[97] Besides studying regions suitable for the accumulation of nodules, the team of marine geoscientists acquired continuous bathymetric, seismic, and magnetic profiles across the Pacific, contributing relevant and unique conclusions about the ocean floor. They found that the fracture zone of Marquesas extended thousands of kilometers underwater, and they published a geological model to predict the formation of polymetallic nodules under different geological conditions.[98]

From 1970 to 1975, CNEXO led a dozen scientific-industrial campaigns across the South Pacific, with the Tahiti base as their point of departure and arrival. However, despite the large investment in nodule prospecting, the Pacific Oceanological Center barely grew. Geologists and geophysicists traveled from Brest to Tahiti, gathered large amounts of data with underwater cameras, dredging techniques, and sampling instruments, and brought them back to Brest, where they analyzed them with microscopes, x-ray mineralogical techniques, and specific instruments to conduct chemical analyses. Conversely, in those initial years, the Pacific base barely employed ten permanent researchers (eight biologists and two technicians), devoted to aquaculture, the biology of tropical fishes, and seafood.[99] As the seventies went by, CNEXO's annual reports gradually ceased to mention the industrial exploration for polymetallic nodules as one of the base's main activities, while aquaculture of crabs and shrimp, or the biology of tuna fisheries, monopolized the center's experiments.

Rather than becoming a large oceanographic nucleus for international research, the Pacific Oceanological Center became a logistical base for

French vessels and researchers embarking on industrial campaigns. In addition to creating an open door to the Pacific, CNEXO was pursuing strategic and geopolitical aims. As la Prairie affirmed, "unlike what happened with Algeria, France would not lose control of French Polynesia and its potential natural resources."[100] Through French government investment in marine research facilities in overseas territories, France was attempting to maintain control over the region in case economically profitable mineral resources existed under French Polynesia's waters.

Meanwhile, information acquired during campaigns across the high seas enabled AFERNOD to estimate the economic prospects of the South Pacific. By 1975, data were sufficient to conclude that the region would not be economically profitable for mining nodules, whereas the North Pacific, already controlled by American and Japanese mining consortia, appeared much more promising. As geologist Michel Hoffert has highlighted in his exhaustive book on France's quest for polymetallic nodules, moving AFERNOD's ventures to the North Pacific entailed long diplomatic negotiations—on some occasions, confidential—with American and Japanese representatives, even signing secret contracts to divide the North Pacific's international waters among them, anticipating the moment when exploiting nodules would be economically profitable and legally possible.[101]

After 1982, international interest in mining deep-sea nodules decreased. The unparalleled scientific and technological development around this industry demonstrated that its extraction would not be feasible for at least a decade. The deep ocean floor was no longer seen as a vast expanse covered by easily recoverable nodules, but a complex geographical region where nodules were scattered in small patches. Globally, industrial growth stagnated and the value of minerals like manganese, copper, and cobalt decreased. AFERNOD continued supporting nodule exploration, promoting varying approaches and strategies depending on the value of minerals and the prevailing international legislation, until it dissolved in 1998.[102] Commercial mining of polymetallic nodules never materialized in the Pacific, and arguably the most remarkable outcome of CNEXO's activities there was the unprecedented geological understanding gained of the seafloor's composition.

DRIVING THE FUTURE OCEAN ECONOMY THROUGH SCIENTIFIC ALLIANCES

In February 1972, France's minister of foreign affairs, Maurice Schumann, wrote a personal letter to Yves la Prairie, congratulating him for the international success CNEXO had rapidly achieved. Schumann acknowledged his deep satisfaction to hear, in high-level diplomatic meetings in Washington,

London, Stockholm, Madrid, or Tokyo, that "thanks to CNEXO, France occupies in oceanology a forefront position, if not the leading one." Schumann highlighted that even Japan's emperor had "spontaneously" started talking to him about the successful Franco-Japanese collaborations in the oceans.[103]

Clearly, reaching this international standing was the result of the intense and well-planned diplomatic strategy CNEXO undertook from its foundation, orchestrated through a network that articulated different governmental bodies, French embassies around the world, the hydrocarbon and mining industries and, as their mediator and public representative, CNEXO itself. Its overarching goal was to establish the most fruitful international collaborations to anticipate the exploitation of the high seas, framed by the ideas, fears, and principles that permeated the international community. As the ocean floor was increasingly perceived as a new global economic frontier, international collaboration became an imperative to prepare for the future exploitation of the oceans. From this perspective, CNEXO's strategy pursued integrating activities of territorial construction and deep-sea technological testing with the international approach to cooperative ocean exploration.

This chapter has focused on two types of diplomatic performances through ocean sciences that enabled France to strengthen its (scientific) position in the Atlantic and across the deep Pacific Ocean. For the first, France found in ocean exploration a means to diplomatically reapproach the US. La Prairie's displays of open admiration for the American organization of ocean sciences, and CNEXO's deliberate attempt to follow that example, symbolized how the goals and views about the oceans of the two countries converged. For the Pacific Ocean, France struck new alliances, like the bilateral cooperative agreement with Japan, and relied on its overseas territories to develop its scientific influence in those waters. Meanwhile, the promise of polymetallic nodules drove the establishment of international scientific relations to survey the ocean floor.

Returning to Minister Schumann's statement prompts a question: What exactly sparked CNEXO's sudden and (seemingly) globally acclaimed success? Had France truly secured a leading role in ocean exploration, or was this merely the perception of a strategically crafted position? The early activities of CNEXO can provide an answer. They reveal that achieving a prominent international position involved more than just scientific accomplishments or developing joint oceanographic missions. It also hinged on the potential research—the ability to conduct significant science with others and the displayed willingness to open resources for collaboration.

CNEXO devoted its first years of existence to nurturing international relations that paved the way for future joint scientific activities. The creation of the two oceanographic centers in Tahiti and in Brest are examples, as well as the research agreements established with other countries like Germany, Brazil, Sweden, the UK, Egypt, or the Soviet Union, among many others. Collaboration in joint expeditions at sea didn't always materialize, but they underscored CNEXO's commitment to lead the exploration of the oceans.

CHAPTER 6

Stories Beneath Deep Salt:
Drilling Across the Mediterranean

Time past and time future
What might have been and what has been
Point to one end, which is always present.

T. S. ELIOT, "BURNT NORTON"

"The floor of the Mediterranean is covered with a thick layer of salt."[1] This was the key takeaway in *Le Monde* after the American-led Deep Sea Drilling Project (DSDP) conducted the most ambitious marine geology expedition yet in the Mediterranean during the summer of 1970. For the first time, this international mission retrieved rock samples more than eight hundred meters below the seafloor, in water depths ranging from a thousand to more than four thousand meters.[2] The samples revealed a thick layer of evaporites—salt—starting 360 meters beneath the seabed. For many readers, then and now, the headline might raise a few questions: Where exactly is this salt? Could we see it, if we sent down a camera? Does it have any commercial value itself? And why should anyone care about salt beneath deep water? Years ago, when I first learned about this Mediterranean expedition—the Deep Sea Drilling Project, its scientific drillship *Glomar Challenger*, and the European scientists who participated—I found myself wondering about these same questions. My fascination with the history of seafloor exploration and the relationship between industry and academia began with this discovery.

Salt in the deep seafloor tells a twofold story: one of Earth's distant past and one of future economic potential. It is a remanent of ancient oceans and linked to potential oil and gas reserves. The seafloor's duality of geological history and commercial resources is why, throughout the period covered in this book, academic geologists and oil industry experts found a common ground for collaboration. To understand the significance of *Le Monde*'s headline, we should revisit our starting point, focusing on how the two main groups involved (French geologists and industry stakeholders) dealt with the seafloor. For Jacques Bourcart, whose work opened this narrative, the seafloor was a vast unknown, a blank space on Earth's maps. His research led him to chart the surface of the French Mediterranean seabed, not far from shore. He died in 1965 without the data needed to probe his theories of a catastrophic transformation of the region—proof that would later be found deep beneath the seabed. In parallel, the emerging French and American offshore oil industries were concentrating on shallow waters. Drawing on land-based models of oil and gas exploration, they focused on detecting formations like salt domes which on land had proved to be a sign of oil deposits. Offshore oil production was limited to the continental shelf; anything beyond was seen as unproductive and, thus, left uncharted.

Beginning in 1968, the DSDP redefined this framework. Equipped with the first drilling vessel designed for scientific research, geologists could finally explore the last *deep* frontier: the rocks beneath the seabed, in the world's deepest waters. The US National Science Foundation (NSF) funded the program, a consortium of American oceanographic institutions organized it, and selected geologists in other countries were invited to join the expeditions. Fragments of rocks and mud collected across the oceans were then analyzed in laboratories around the globe. As geoscientists from different regions shared their insights, the vastness of the ocean floor seemed to shrink. The DSDP expeditions connected distant geographies by drawing on parallels, models, and equivalent formations, revealing patterns of crustal movement, past environmental changes, and the evolution of ocean basins over time.[3] And, for the Mediterranean, the DSDP discovered a massive salt layer: an intriguing setting for scientists studying past environmental catastrophes, and a promising area for French industry stakeholders.

This American big-science project, its technology, and its results are an integral part of France's commercial, environmental, and scientific history—particularly regarding its submerged territory. Samples from two distant geographies yielded two unintended yet crucial insights that shaped France's vision of its oceanic future. First, exploration in the Gulf of Mexico provided a model for deep-water oil and gas discoveries, which could be applied to the Mediterranean basin. Second, it brought to the western Mediterranean, considered France's "backyard," a unique tool for seafloor study: the scientific drilling vessel. This exploration model, combined with the new samples it provided, transformed the deep Mediterranean into an area of both energy potential and geological interest. This chapter focuses on the 1970s, when French industry stakeholders came to see the Mediterranean as Europe's equivalent of the Gulf of Mexico—a resource-rich, deeply submerged territory beyond the shallow continental shelf. Just as oil from the Gulf of Mexico fueled American society, the western Mediterranean held the potential to power France's projected oceanic economy. This view spurred both geologists and industry to collaborate internationally, deepening their joint understanding of the area.

Beyond the significance of these events for France's views on its underwater territory, the Mediterranean DSDP expeditions represented an experiment in international collaboration that transcends conventional academic frameworks: French industry experts participated as unofficial scientific partners. In this way, this chapter follows the previous ones by examining one of the modes in which industry, science, and international relations came together in the study of the seafloor, producing France's unique perspective in the global oceanic order. However, one key distinction sets France's involvement in the DSDP apart from the national undertakings so far discussed: in this case, France played a secondary role. French funding agencies did not contribute economically to the program. French geologists did not lead or organize the cruises. The French scientific delegation was simply an additional international participant. And yet this story showcases how CNEXO and France's oil industry organized to make the most of these international surveys— specifically, gathering geological information in areas where, due to economic or technical constraints, commercial surveys would not be conducted in the short term. International science and national industry intersected to deepen their understanding of the deep Mediterranean seafloor. The nature of the underwater environment, this new frontier of depth, required such collaboration.

One final reason the DSDP Mediterranean expedition merits attention: there is already a narrative about it. Geologist Kenneth J. Hsü, one of the organizers and co–chief scientists of the cruise, published in 1983 *The Mediterranean Was a Desert: A Voyage of the Glomar Challenger*. This firsthand account has shaped the scientific narrative about the discovery of the Mediterranean's deep geology, becoming the staple reference to understand the sequence of events, the motivations, the decision-making processes, and the different stances in the scientific debate that unfolded around the origin of the Mediterranean salt domes. However, Hsü's narrative dismisses oil exploration as "a triviality" and leaves it out entirely.[4] In this chapter I argue that oil exploration was far from trivial; it was crucial in many ways to one of the Mediterranean's most significant geological discoveries. France's oil industry, in particular, not only played a central role in studying deep salt domes but also contributed significantly to the planning of the DSDP cruises. This chapter closes the book by showing how the oil industry advanced scientific knowledge in areas important to the scientific community. Recognizing the oil industry's contributions and understanding how they shaped knowledge production allows us to build a more nuanced view of what we know about the seafloor today.

DRILLING BEYOND WATER, FROM FIXED PLATFORMS TO FLOATING LABS

Look closely at the image of a drillship and you will see an awkward hybrid: it resembles a cargo vessel with an onshore oil rig mounted atop. It is a technology that brings together the realms of seafaring and drilling operations with the purpose of exploring the ground beneath the fluid, marine environment. A drillship's key feature is its mobility, a quality that wasn't always essential in the early offshore oil industry. The need for mobility only arose with the seafloor's exploration rush, as oil companies (and marine geologists) sought to expand into open waters.

In the late nineteenth century, as American oil firms began experimenting with drilling in muddy regions and shallow coastal waters, they developed methods to bring their onshore technological setup to the fluid environment. On Ohio's lakes, derricks stood on platforms built over piles. In Louisiana's wetlands, timber mats assembled into grids provided a stable base for installing derricks and other equipment.[5] Along California's coastline, wooden piers with derricks on top extended up to four hundred meters from the beaches of Santa Barbara.[6] At the turn of the century, on Louisiana's Caddo Lake, oil operators pioneered drilling from floating platforms such as barges and tugboats. In all these cases, water depth did not exceed a dozen meters.[7]

The need for greater mobility arose at the onset of World War II, as oil demand intensified and American companies turned their attention to areas beyond the coastline. In 1947, the platform Kermac 16 represented the first attempt to disconnect oil operations from the shore. Towed to its position 1,700 meters off Louisiana's Gulf coast, over five meters of water, it became the first facility out of sight of the coastline. Modified US Navy vessels served as supply boats, transporting crew and equipment in and out.[8]

This platform laid the groundwork for the industry's next step: moving from towed platforms to mobile drilling units—vessels equipped with a complete drilling rig and its ancillary gear, including a mud-handling system, a drill pipe, and a system of well-cementing.[9] In 1956 a consortium formed by the oil companies Continental, Union, Shell, and Superior Oil financed the design and construction of the first drillship, *CUSS I*, its name an acronym for the group. Its mission was to explore the oil potential of the California continental shelf beyond the shoreline. After preliminary tests, the *CUSS I* made history when its crew obtained a core sample from a well at a water depth of 122 meters, the deepest offshore well drilled to that date. *CUSS I* would not have been able to operate far from the coastline without its innovative dynamic positioning system. Drilling operations in the high seas require the vessel to hold a stable position for hours or even days. As long sections of drill pipe are suspended between the drillship and the ocean floor, any sudden movement could break the pipe. The first dynamic positioning system used a circular array of sonar buoys that sent signals to a control booth on the vessel. The drillship's position was then manually adjusted using four thrusters, one at each corner of the vessel.

CUSS I was soon modified from oil exploration work to scientific research, manifesting the establishment of a tight relationship between oil and scientific ocean drilling. In 1958, the vessel was put at the service of the Mohole Project, an ambitious American scientific program to reach and sample the Mohorovičić discontinuity (the boundary between Earth's crust and mantle).[10] Organized by the American Miscellaneous Society, a group of American oceanographers and Earth scientists, the Mohole Project epitomized the scientific race during the Cold War: American experts sought to surpass the Soviets in being the first to reach Earth's mantle, a competition akin to the one established to reach outer space. Indeed, the Soviet Union was in parallel launching a scientific (continental) drilling program to sample the mantle by drilling in the Kola Peninsula.[11] Although drilling on land was much more common and logistically easier, American organizers decided to drill through the oceanic crust because it is notably thinner than the continental crust (an average of 10 kilometers thick in the oceans, versus 40 on the continents). Operations started in March 1961, drilling five sites off Guadalupe Island, 400 kilometers south of San Diego. After that first attempt, the program came to a halt due to administrative conflicts, escalating costs, and technical problems. After eight years, no new drilling, and 57 million dollars provided by the NSF, the Mohole Project was canceled in 1966 without moving out of its preliminary phase.

Although many considered the project a failure, it set a scientific milestone by demonstrating that technological capabilities to drill through great depths of water and ground existed: Mohole's deepest well reached 183 meters below the seafloor, in 3,600 meters of water. Equipped with its cutting-edge dynamic positioning system, it was the first time an untethered drilling platform could stay steady in such depths.[12] Conscious of the great scientific potential of deep-sea drilling technologies, the American organizers conceived Mohole's successor: the Deep Sea Drilling Project (DSDP).

The DSDP was an American-led, international scientific project aimed at understanding the geology and evolution of Earth. A brand-new drillship adapted to the requirements of scientific sampling would traverse the world oceans, collecting rock and sediment samples from the deepest regions of the ocean floor. The NSF supported the DSDP economically, signing a contract in June 1966 for 12.5 million dollars with the Scripps Institution of Oceanography, the DSDP main operator. Scripps awarded the design and construction of the scientific drillship to the American oil-tech company

Figure 15. The first scientific deep sea drilling vessel, the *Glomar Challenger*. Launched in 1967, the Scripps Institution of Oceanography managed its operation under contract to the National Science Foundation. Here pictured in the eastern equatorial Pacific Ocean in April 1982, during DSDP Leg 85. Source: Special Collections and Archives, UC San Diego, La Jolla. Reprinted with permission of UCSD.

Glomar Marine Inc., which specialized in designing floating rigs for off-shore oil firms and devices for deep-sea mining.[13]

The D/V *Glomar Challenger* (fig. 15) was named as a tribute to the famous nineteenth-century oceanographic vessel HMS *Challenger*. Its design was based on the latest industrial innovations: Its hull drew from Glomar Marine's latest and most advanced drillship, the D/V *Grand Isle*, modified to operate in unprecedentedly deep waters. For the *Glomar Challenger*, the firm's engineers designed a revolutionary dynamic positioning system that did not require manual intervention. Four transponder beacons lowered to the seabed reported the vessel's position, activating four computer-controlled propellers installed in the hull to correct it. The *Glomar Challenger* was also the first commercial vessel that relied on the US Navy's satellite navigation and communication system.[14] Other innovations, already common in Glomar Marine drillships, increased the efficiency of scientific drilling. A reentry system allowed accurate return to a previously drilled hole; the drill pipe could extend up to seven kilometers through the seawater and the seafloor; and the derrick towered forty-three meters above the drilling platform. The higher the derrick, the more efficient the platform becomes for drilling deeper, because tall derricks allow for longer pipes to be used without the need to add new sections, thus speeding up the drilling process.[15]

With the *Glomar Challenger*, marine geology entered in the realm of big science, and not only because of the massive dimensions of the vessel, a hundred twenty meters long and twenty meters wide. The drillship was a floating laboratory that required several groups of experts to operate. Its cruises were not cheap: at that time, one day at sea cost about 30,000 US dollars (roughly 278,000 USD in 2025). To that cost were added other expenses derived from maintenance and repairs, laboratory equipment and supplies, and technical improvements undertaken when the ship was in port.[16] In March 1968, Glomar Marine turned the vessel over to the Scripps Institution of Oceanography and, after a shakedown run and some fine-tuning, its first cruise inaugurated almost six decades of scientific ocean drilling around the world.

UNEXPECTED SCIENCE: A NEW GLOBAL MODEL FOR DEEP OIL

In the summer of 1968, the *Glomar Challenger* departed from Orange, Texas, to venture into the Gulf of Mexico for its first scientific cruise, known as Leg 1.[17] The region became the testing ground for the scientific drillship for both commercial and scientific reasons. Its waters were well known to American offshore operators: The Gulf of Mexico was the most explored,

Table 6.1. Exploratory activity in the Gulf of Mexico's continental shelf

Year	Exploratory wells	Proved fields	Average water depth
1947–59	442	70	16
1960–69	2,377	119	36
1970–79	3,356	265	59

Source: Modified after Lore, "An Exploration and Discovery Model."

drilled, and developed offshore oil province in the world.[18] Since 1947, American oil firms had drilled almost three thousand exploratory wells at an average water depth of thirty-six meters, which had enabled engineers, technicians, and oil geologists to develop vast technical expertise in studying marine sediments and in mitigating the risks associated with underwater drilling (table 6.1).[19]

Besides its industrial significance, the Gulf of Mexico had also been the object of numerous scientific geophysical cruises during the 1950s and the 1960s, most of them led by American geophysicist Maurice Ewing, from Lamont. As detailed in chapter 3, Ewing had pioneered the use of marine geophysical techniques to understand the structure and composition of the ocean crust. After surveying the Gulf of Mexico, he and his team grew intrigued by the dome-like structures that appeared in seismic profiles of the Gulf's abyssal plain. If those were salt domes and not geophysical artifacts, their likely existence below 3,000 meters of water challenged current knowledge about salt dome formations.[20] As salt deposits accumulate after the retreat and evaporation of seawater, geologists considered that, in the marine domain, they could only be found in shallow areas resulting from past marine regressions over the continental shelf. Ewing's investigation was particularly relevant for the oil industry. As oil geologists working on land had discovered in the early twentieth century, salt domes and layers constitute excellent traps for hydrocarbon deposits, yet their existence beyond the nearshore had never been confirmed. If the existence of salt domes in deep waters was proved, it would imply that oil and gas reservoirs could be found there. The DSDP offered a unique opportunity to probe Ewing's hypothesis by drilling in the Gulf's abyssal plain, in the region called Sigsbee Knolls.[21]

Ewing, already in his sixties, became the co-chief of the scientific party during the *Glomar Challenger*'s initial expedition of 1968, together with his colleague from Lamont, the geologist J. Lamar Worzel. While testing the drillship's drilling gear, the expedition aimed at advancing knowledge of the Gulf's geological history and its dome-like formations. Non-US scientists would join later expeditions as the *Glomar Challenger* sailed away from

American waters. Yet, for this first cruise across the Gulf of Mexico, only experts from US institutions were onboard. The composition of the first team in the DSDP reflects the merging of expertise required to run a scientific drillship: the scientific skillset was contributed by three paleontologists (two from research institutions and one from the US Geological Survey) and three geologists—two subcontracted from oil companies to work in the DSDP as staff scientists for Scripps. A team from Global Marine Inc. was responsible for the ship's drilling operations and navigation, while lab technicians and engineers from Scripps ensured the functioning of the onboard laboratory facilities. Two representatives from the NSF and two photographers from Scripps completed the team.

The expedition started off successfully. The crew drilled and recovered continuous sections of sediment from the Gulf's deepest region, beneath more than five thousand meters of water—deeper than any oil company had ever ventured.[22] But a week into the expedition, a finding marked a turning point no one had anticipated: while drilling at thirty-five hundred meters deep in the Sigsbee Knolls region, a retrieved core came out spilling oil (fig. 16). The drill pipe had hit a cap rock, a geological structure that covers and seals hydrocarbon deposits. The drilling was immediately stopped and the borehole filled and abandoned to prevent an oil spill.[23]

Although the cruise's main goal was scientific, Leg 1 of the DSDP is remembered for its commercial significance: it demonstrated that oil deposits could form at depths previously unsuspected. Finding hydrocarbon deposits around salt domes was commonplace, but occurrences at such depths altered all principles of hydrocarbon exploration. At the same time, the findings in the Sigsbee Knolls proved that the technological capabilities to access potential deep-sea deposits already existed, paving the way for the oil industry to venture into this new frontier.

The Gulf of Mexico's abyssal plain could hide oil-rich salt domes and, following this logic, so could any other deep basin with salt domes or evaporitic formations in its seafloor composition. Accordingly, at the other side of the Atlantic, the Mediterranean basin emerged as an ideal candidate. On September 4, 1968, the French newspaper *Le Monde* announced with fanfare that, during the DSDP Leg 1, American geologists had found hints of oil deposits for the first time in deep waters.[24] The finding had a major impact within CNEXO's leadership.[25] During a meeting of CNEXO's Administrative Council, André Giraud (representing France's oil industry) expressed once again his fear of finding themselves "highly distanced from the Americans," but this time in the domain of deep-sea drilling. He was concerned that the US could use the *Glomar Challenger* for oil exploration disguised under the label of "scientific exploration," especially if the

Figure 16. The co-chief scientists of DSDP Leg 1, Maurice Ewing (left) and J. Lamar
Worzel (center), looking at an oil-laden sediment core recovered in the region of
Sigsbee Knolls, in the Gulf of Mexico. The sample was retrieved below more than 3.5 ki-
lometers of water, 146 meters below the seafloor. On the right, Jim Dean, from Mobile
Oil and cruise operations manager aboard the *Glomar Challenger*, is holding the core's
plastic container. Source: Special Collections and Archives, UC San Diego, La Jolla.
Reprinted with permission of UCSD.

drillship was going to sail near submerged French territories.[26] The DSDP had been launched as a scientific program and at that time only American-based scientists participated. So, what if American oil companies became involved in organizing the expeditions? What would happen if the *Glomar Challenger* progressively transformed into a support platform for oil exploration in deep waters?

Giraud was concerned about the western Mediterranean. The geological formations that Ewing's team had found in the deep Gulf of Mexico seemed very similar to the structures that a few years earlier, in 1965, American oceanographers John Hersey and Henry Menard had identified in the deep western Mediterranean (as we saw in chap. 2). If these were also salt domes, the Mediterranean would become a promising oil field, close to mainland France and up for grabs. However, what troubled Giraud most was that French oil companies lacked the technology to explore or drill the abyssal plain—only the *Glomar Challenger* could do it. The French Petroleum Company (CFP) had just drilled its first borehole in the Gulf of Lion's continental shelf, in waters shallower than a hundred meters deep, by leasing the D/V *Drillship* from the American firm Drillship Associates.[27] In the mind of CNEXO's leadership, it would be evidently regrettable if the *Glomar Challenger* was sent into the Mediterranean—a region understood by French oil companies as their backyard—before French academic or industrial experts possessed detailed information about its economic potential or geological value. That could pass the advantage to the Americans in future projects to exploit its richness.[28]

Giraud's worry about finding the Americans drilling into the deep Mediterranean was not totally unfounded. While these discussions unfolded at CNEXO's headquarters in Paris, American oceanographers across the Atlantic began planning a DSDP cruise in the Mediterranean. This expedition, part of the drillship's round-the-world tour, was not targeting potential deep reservoirs, but sought to study the basin's geological significance within the framework of plate tectonics.

THE STRATEGIC DIMENSIONS OF SCIENTIFIC OCEAN DRILLING

To understand why this American program ventured into the Mediterranean early on, we must first examine how its expeditions were organized. The DSDP set out to explore the deep oceanic crust on a global scale by conducting two-month-long, back-to-back expeditions. Among its top scientific priorities were confirming and deepening the study of plate tectonics, which required a worldwide survey of the ocean floor. After Leg 1 in the

Gulf of Mexico, the program continued with a transatlantic expedition from Hoboken, New Jersey, to Dakar, Senegal, heading then to Rio de Janeiro, Brazil. It was this third expedition, Leg 3, that provided critical confirmation of plate tectonics. Core samples retrieved from the Mid-Atlantic ridge revealed patterns of magnetic striping that aligned with the predictions for seafloor spreading. The *Glomar Challenger*'s global tour then continued north to the Panama Canal and California, across the Pacific to Guam via Hawaii, and onward to French Polynesia.

The schedule for expeditions was planned about year in advance. The endpoint of one expedition served as the starting point for the next, optimizing time, money, and effort. During brief port calls, the crew and the scientific team were rotated, sediment samples were shipped to the Scripps Institution of Oceanography, the vessel was resupplied, and the new expedition began. During the first dozen DSDP expeditions, most scientists onboard were based at American institutions. Of ten to twelve geologists per voyage, only a maximum of three were invited from foreign research centers (table 6.2).

For 1970, the American organizers of the DSDP planned an eastward route: from Tahiti to Panama, through the canal to Florida, and back into the North Atlantic Ocean. It came as a natural follow-up to head into the Mediterranean Sea, a region with major historical, cultural, and geological significance. It was in spring 1969 that CNEXO experts first heard of plans for a Mediterranean DSDP expedition. The word came through informal channels: the American-based proponents, William B. F. Ryan and Kenneth J. Hsü, were getting in touch with Europe-based colleagues to help them lay out a drilling plan in the Mediterranean basin. This led them to marine geologist Xavier le Pichon, by then scientific advisor to the general director at CNEXO. Ryan, a young marine geophysicist from Columbia University's Lamont, was one of the leading US experts on the western Mediterranean seafloor despite being under thirty. He began his career in 1961 under Maurice Ewing and Bruce Heezen, pioneers in marine geophysics and continental drift studies. For his PhD, Ryan analyzed data from Ewing's late-1950s Mediterranean cruises, producing the first comprehensive tectonic overview of the basin and a detailed study of its dome-like formations.[29] Accompanying Ryan as coproponent of the Mediterranean expedition was Kenneth J. Hsü, a Chinese geologist who had developed his scientific career in American research institutions. Although he moved to the Swiss Federal Institution of Technology in Zurich when the DSDP started, he was among the first onboard participants in the *Glomar Challenger*.

The need to involve European colleagues in a *Glomar Challenger* expedition stemmed from both the regional expertise of their research teams

Table 6.2. Details of the initial cruises of the Deep Sea Drilling Project

Expedition (dates)	Route	Country of affiliation of onboard scientists (number)
Leg 1	Orange, TX to Hoboken, NJ (USA)	USA (8)
Leg 2 (Oct–Nov 1968)	Hoboken to Dakar (Senegal)	USA (9), Italy (1)
Leg 3 (Dec–Jan 1969)	Dakar to Rio de Janeiro (Brazil)	USA (8), Switzerland (1)
Leg 4 (Feb–Mar 1969)	Rio de Janeiro to San Cristobal (Panama)	USA (8), Switzerland (1)
Leg 5 (Apr–June 1969)	San Diego, CA to Honolulu, HI (USA)	USA (8)
Leg 6 (Jun–Aug 1969)	from Honolulu to Apra (Guam)	USA (9), USSR (2)
Leg 7 (Aug–Sept 1969)	from Apra to Honolulu	USA (8), Switzerland (1), West Germany (1)
Leg 8 (Oct–Dec 1969)	Honolulu to Papeete (French Polynesia)	USA (6), Switzerland (1), France (1), Sweden (1)
Leg 9 (Dec–Jan1970)	Papeete to Balboa (Panama)	USA (9), New Zealand (1)
Leg 10 (Feb–Apr 1970)	Galveston, TX to Miami, FL (USA)	USA (9)
Leg 11 (Apr–June 1970)	Miami to Hoboken	USA (8), France (2)
Leg 12 (Jun–Aug 1970)	Boston, MS to Lisbon (Portugal)	USA (5), UK (2), Canada (2), Germany (1), Denmark (1),
Leg 13 (Aug–Oct 1970)	Lisbon to Lisbon (roundtrip across the Mediterranean Sea)	USA (1), Switzerland (2), Italy (2), France (2), Romania (1), UK (1), Austria (1)

The name of the country and US state are only mentioned the first time a location appears.
Source: Based on the Initial Reports of the Deep Sea Drilling Project, vols. 1–13; table by the author.

and the diplomatic intricacies of the seafloor—or, as Hsü summarized with a touch of irony in his book about this Mediterranean cruise, "it would be discourteous to European colleagues if Americans were to poke holes in their backyard without their active participation."[30] There were wide-ranging scientific motives for having non-US researchers engaged in the planning phase and onboard. For planning the cruise, Ryan and Hsü needed detailed data on the seafloor's composition. Geophysical profiles would help them identify suitable drill sites where the sediment sequence met their expectations, and where they could avoid cracks or faults that might disrupt drilling operations. Although American institutions stored numerous geophysical profiles from the Mediterranean seafloor (acquired during previous around-the-world campaigns), those were not sufficiently detailed to organize a drilling cruise across the basin. At the same time, European-based research teams were a pool of scientific expertise. Their understanding of the geological formations was fundamental to analyzing and interpreting the drilled samples, while the huge amount of data produced during each *Glomar Challenger* campaign required a combined effort that was beyond the capabilities of American researchers.

Moreover, the program's decision to invite non-US experts stemmed from recent concerns regarding the scientific control of the high seas. By the late 1960s, as we saw earlier, UNCLOS delegates from emergent nations expressed growing concern about research activities conducted across the high seas, especially those that had obvious commercial connotations and were carried out by nations leading oceanographical research. The DSDP fell firmly into this category. Therefore, although the program was totally subsidized by the American NSF, it was important to signal its international, purely scientific nature. The presence of foreign researchers onboard prevented suspicions among foreign authorities about possible commercial goals hidden behind the Americans' pursuit of drilling in international waters around the world. For example, French researchers had been invited earlier to board the drillship because the cruise would end up in Papeete, French Polynesia. At that time, sedimentologist Wladimir Nesteroff, from the University of Paris, was invited because of his previous collaborations with American oceanographers at the Scripps Institution of Oceanography.[31] These kinds of diplomatic gestures were well encapsulated in a letter la Prairie wrote to the governor of French Polynesia before the *Glomar Challenger* reached port, describing the scientific and political importance of the DSDP and cordially encouraging him to receive the *Glomar Challenger*'s crew "with the best traditions of your legendary hospitality."[32] Building on this diplomatic groundwork, we can now refocus on the moment when Ryan, one of the Mediterranean cruise organizers, reached out to Xavier le Pichon at France's CNEXO.

Ryan and le Pichon were close friends from the period le Pichon had spent conducting research at Lamont, from 1963 to 1968.[33] Ryan was probably aware of CNEXO's research means and, when he got in touch with le Pichon, he explained their plans to conduct a deep-sea drilling survey across the Mediterranean. The expedition's key interest was to understand the origin of the basin within the mounting evidence for plate tectonics. Until this time, geologists had considered the Mediterranean basin a depression of Earth's crust. However, from the new perspective of plate tectonics, the Mediterranean became a sort of seam between the African and Eurasian plates.[34] Only deep-sea cores, recovered with the *Glomar Challenger*, could throw light on the Mediterranean basin's age, origin, and dynamics.

Studying the dome-like structures of the deep basin was one of the secondary objectives of the expedition. As noted earlier, these structures had been identified in the Mediterranean through seismic profiles by American geoscientists Hersey and Menard.[35] The fact that the potential salt domes were more intriguing for industry stakeholders than for scientists is especially significant, as it reveals the particular ways in which industry and scientific interests intersected. For geologists, the key focus was plate tectonics. Determining whether the dome-like formations were composed of salt, like those in the Gulf of Mexico, or whether they were something else—such as buried submarine volcanoes—was crucial only if that could add to the understanding of the region's geological processes, and in a broader scale, to the evolution of ocean basins. The expedition was expected to produce high-impact scientific results about the tectonics of the region. For Xavier le Pichon, then, the key motivation behind involving French researchers in the DSDP Mediterranean cruise was the opportunity to elevate French marine geology on the international stage, rather than the discovery of deep hydrocarbon reservoirs. Immediately after hearing from Ryan, le Pichon urged CNEXO's General Director Yves la Prairie to get his research team involved in the campaign's organization as soon as possible.[36] It is likely that France's oil industry experts learned about the planning of the Mediterranean expedition through these channels and became interested in how they could be involved.

Providing high-quality data to the expedition organizers would translate into a strategic advantage in shaping the expedition. Drill sites could only be chosen based on the available geophysical data. For both CNEXO researchers and French oil industry experts, supplying data that aligned with the expedition's scientific goals increased the likelihood that those locations would be selected for drilling. Hsü explains in his 1983 account of the expedition that as soon as he started planning it, in October 1969, "industry people in France" sent drilling proposals (presumably, suggesting conducive drill sites) and asked him for an appointment. Hsü doesn't delve into his response to them, but French industry experts were invited to a planning meeting held one month later in Zurich. Several oceanographic institutions in Europe also sent representatives, including CNEXO. Le Pichon played a key role alongside Hsü in that meeting, though Hsü's account remains the only source on what occurred. Hsü recalls the pressure he faced from French oil experts, who believed the drilling vessel could be used for oil exploration and urged drilling on top of salt domes (as DSDP Leg 1 had accidentally done in the Gulf of Mexico). As Hsü recalls, le Pichon "rescued me from the siege by his countrymen, pointing out that the purpose of the deep-sea drilling was to investigate the origin of ocean basins, not to get lost in trivialities."[37] Hsü's narrative sheds light on two key points: first, how the scientific story has sidelined intense commercial interests; and second, the presence behind the scenes of France's oil industry within this international scientific project.

Once we understand the entanglements between science and industry at CNEXO, it seems plausible that France's oil industry found another way to participate in suggesting the drill sites. In March 1970, five months before the Mediterranean expedition started, CNEXO undertook a series of campaigns aimed at recovering geophysical and sedimentological information to suggest drill sites.[38] These new datasets augmented a wealth of seismic profiles acquired during earlier surveys, together with sedimentological information gathered during the industrial-academic campaigns described in previous chapters. Based on CNEXO's data, Hsü and Ryan selected three drill sites: in the salt dome region of the Balearic basin (site 124), in the Sardinian Channel (site 133), and over the western Sardinian continental slope (site 134).[39]

Let's take a moment to closely examine the oil industry's role in this planning phase. It appears that, through CNEXO, France's oil industry contributed data to help characterize the regions for sampling. However, there is room to question whether there was a deliberate attempt to steer drilling with the aim of identifying oil and gas reservoirs—particularly drilling across cap rocks, as Hsü's narrative suggests. The *Glomar Challenger* lacked the safety systems required to drill into gas or oil deposits, and it seems unlikely that oil geologists would advocate for high-risk operations.[40] Piercing into such highly pressurized reservoirs could have resulted in a catastrophic explosion and massive oil spill. By the 1970s, the oil industry was acutely aware of these risks, especially after two major accidents: the sinking of the *SS Torrey Canyon* in March 1967, and the explosion of a Union Oil platform in Santa Barbara in January 1969. These disasters underscored the importance of safety standards and heightened security around drilling near potential hydrocarbon deposits. Moreover, the DSDP Executive Committee included an Advisory Panel on Pollution Prevention and Safety, responsible for evaluating the safety of every proposed drilling site.[41] This panel, composed of a dozen experts from US universities and private companies, explicitly prohibited drilling near or over salt domes, where hydrocarbons were most likely to be found.[42]

So, how could CNEXO's suggested drill sites benefit French oil companies, if the Mediterranean expedition was not directly targeting oil deposits? The answer highlights the strategic nature of these early academic-industrial collaborations in seafloor exploration. For industry stakeholders, the value of drilling deep in the Mediterranean seafloor and investigating the enigmatic salt domes did not lie in immediately discovering oil reservoirs. Instead, the goal was to obtain detailed samples of the deep basin, ground-truth geophysical data, and create a clearer picture—both figuratively and literally—of what lay beneath the water. Determining the age of the salt domes and the surrounding formations was critical to assessing their hydrocarbon potential, maturation stage, and quality. Any information about the salt domes' age or the environment in which the salt layers were deposited was greatly valuable.

A SALTY FRENZY: FRENCH INDUSTRY AND INTERNATIONAL SCIENCE AT PLAY

At daybreak of August 18, 1970, the drilling vessel *Glomar Challenger* crossed the Strait of Gibraltar to venture for the first time into the Mediterranean with a science crew of nine onboard. Two French experts embarked, in recognition by the US organizers for contributing to the organization of the cruise as well as for their experience in studying the Mediterranean seafloor. After drilling its first successful borehole in the Alboran Sea, the *Glomar Challenger* headed toward the salt dome region in the Balearic basin. There the team recovered the first deep-sea core that displayed a complete sequence of evaporitic materials: gypsum, halite, and anhydrite (fig. 17). The discovery confirmed the hypothesis of potential deep hydrocarbon reservoirs in the western Mediterranean. As we've seen, salt in the deep sea tells a compelling story about Earth's past and holds a tantalizing promise for the future. The existence of salt domes in the deep Mediterranean set the stage for collaboration between academia and industry that transcended France's boundaries. Two factors acted as catalysts: new spaces of collaboration, partially created by CNEXO, and an increase in oil prices that made the deep seafloor economically feasible.

The scientific narrative of the expedition highlights the discovery that the Mediterranean may have dried up in the ancient past. In the onboard laboratory, Italian micropaleontologist Maria Bianca Cita dated the salt deposits, revealing they had formed relatively recently (5 to 9 million years ago), not during the basin's initial formation.[43] Cores also showed oceanic sediments above the salt, suggesting a rapid transition from a brackish to an open-water environment. Hsü proposed that a sudden disruption of Atlantic inflow led

Figure 17. Kenneth J. Hsü (left), William Ryan (right), and Charles Simon (center, of the *Glomar Challenger*'s drilling crew), holding a just-retrieved core with evaporites during DSDP Leg 13, the Mediterranean Expedition. The evaporite appears at the bottom of the core liner as a fragmented group of whitish salt rocks. Source: Special Collections and Archives, UC San Diego, La Jolla. Reprinted with permission of UCSD.

to massive evaporation, causing the Mediterranean to become a desert.[44] This theory aligned with Jacques Bourcart's earlier findings of seawater regression on France's continental shelf. Hsü's hypothesis—not supported by all the onboard scientists—described a desert-like Mediterranean with brackish lakes and marshes during what became known as the Messinian Salinity Crisis.[45] The event sparked a decades-long geological debate, with experts divided into three main positions. There was common agreement that tectonic movements restricted water inflow to the Mediterranean, but opinions differed on how the salt deposits formed: Hsü's hypothesis of a near-total desiccation in a deep basin; an alternative, suggesting that the salt layer formed in very salty, deep water; and a third envisioning shallow, salty lagoons periodically refilled from the Atlantic.[46]

While this debate unfolded onboard and on land, the archival documents offer a parallel narrative, suggesting a commercial fervor for the deep Mediterranean following the DSDP expedition. Two days after the *Glomar Challenger* moored back at Lisbon on October 7, 1970, la Prairie organized an international press conference sponsored by CNEXO. Journalists, researchers, and oil industry experts from all over the world, as well as the American leaders of the DSDP, attended to hear about the cruise's preliminary results.[47] Hsü, Cita, and Ryan offered detailed accounts of their findings. The compelling image of a landscape so different in a region so familiar triggered astonishment among the audience. For journalists, the imaginary of a desertic basin inspired gripping headlines. *Le Monde* announced that "The Mediterranean seafloor is covered by a thick layer of salt" (which was misleading, as the evaporitic layer was buried below hundreds of meters of more recent sediments), while the journal *Scientific American* published a paper by Hsü entitled "When the Mediterranean Dried Up."[48] And, for the oil industry, the hypothesis of potential salt domes was now grounded on physical evidence.

For French oil companies, it was now clear that the Mediterranean could become "for political and technical reasons" one of the first regions in the world where producing deep hydrocarbon deposits was feasible.[49] After the press conference, oil activities and scientific exploration skyrocketed across the Mediterranean basin. Between January 1971 and January 1973, French and foreign oil companies requested twenty exploration permits from the French Ministry of Industry across the western Mediterranean, 80 percent of them beyond the continental shelf and, of those, half at more than 2,000 meters deep.[50] At the same time, investigations of the giant Messinian salt deposit, its geological origins, and the quantity of hydrocarbons it could harbor transformed the exploration of the deep Mediterranean into a fashionable research subject, while debates in parallel on plate tectonics placed the Mediterranean under great geological scrutiny.[51]

Plate tectonics offered a new geological framework through which oil occurrences might be predicted in the offshore domain. Oil matures in sedimentary basins, convex regions favorable to the accumulation of sediments, usually formed near continental margins. Since the motion of tectonic plates could explain how, where, and when continental margins and sedimentary basins were formed and destroyed, it could offer a powerful approach to oil prospecting, suggesting guidelines for identifying promising regions to explore and even to predict oil and gas occurrences.[52] For the Mediterranean this was a problematic change of perspective: traditionally, the region had been considered a depressed region in permanent subduction where sediments tended to accumulate.[53] If the Mediterranean was not a depressed and stable basin, but the seam between the Eurasian and African plate, all existing data needed to be reinterpreted.

The converging interests of the oil industry and academic geologists stimulated exchanges between the two communities, prompting the oil industry to share its data. One reason for this was rooted in the particular function CNEXO had supported since its 1967 founding, that is, deploying national efforts in marine scientific research to speed up and facilitate industrial exploitation of marine resources. Hence, when French oil companies manifested their interest in exploring the Mediterranean's depths, CNEXO launched research projects that connected, in different ways, three communities: CEPM, the network of national oil companies under the leadership of the Ministry of Industry; the French Institute of Petroleum (IFP), a public institution devoted to offering scientific and technical support to CEPM; and research centers, including geologists attached to CNEXO and to public laboratories.[54]

For oil companies, relying on the support of academic geologists was beneficial for both scientific and economic reasons. Studies in fundamental geology were essential to developing theoretical models on the chances of finding oil deposits in different conditions and structures, while regional studies were useful to assess the oil potential of particular areas. Conversely, geological exploration by oil companies was around ten times more expensive than subsidizing academic research: to lease a foreign drillship, to outsource private companies for preliminary geophysical surveys, or to mobilize petroleum geologists was more costly than supporting scientific missions, outsourcing academic geologists, or coordinating cooperative studies through CNEXO.[55] In this framework, petroleum geologists formalized collaborations with academic geoscientists, orchestrated under CNEXO's sponsorship. One particular example is the Mediterranean Project, launched early in 1971. It aimed at compiling the first geological synthesis of the Mediterranean basin by bringing together petroleum and academic geologists under CNEXO's leadership. The oil industry's interest in moving oil exploration beyond the continental shelf shaped the resulting report: information contained would facilitate the identification of promising oil regions in deep waters, the first step before acquiring exploration leases.

Given that the main aim of the oil industry was to identify favorable regions for hydrocarbon accumulation while reducing time and costs, it also appeared as a suitable strategy to collaborate with the international scientific community. In 1972, experts from French oil companies began attending the biannual congresses of the International Commission for the Scientific Exploration of the Mediterranean (CIESM) in notable numbers.[56] Scientific talks, like those presenting the preliminary results of the cores retrieved during the DSDP's Mediterranean cruise, were as informative for petroleum experts as for the academic attendants.[57] Conversely, petroleum geologists disclosed data and results obtained during oil prospecting activities. In these meetings, new hypotheses and results were presented and debated, building knowledge about the Mediterranean's evolution, structure, and history. For petroleum geologists, being actively involved in the academic community helped them to better interpret their results, keep up with the latest advances and debates, and meet experts with whom to establish future collaborations.

COMMERCIAL DATA FOR A DEEPER SCIENCE, SCIENTIFIC TECHNOLOGIES FOR DEEPER OIL

Multiple meetings and accumulation of data about the formation of the Mediterranean salt domes did not resolve the scientific debate; instead, they fueled it with additional evidence. The main issue was that Leg 13 had drilled only shallow boreholes, retrieving samples from above the salt layer. In 1972, Hsü led the organization of a second DSDP campaign across the Mediterranean, focused on understanding the ancient conditions in which the Mediterranean salts formed. To access that natural environment, they needed to drill much deeper than in the first cruise. As already mentioned, drilling through the salt layer was not an option due to safety concerns. So, how to gain access to the "world" hidden beneath the salt? The solution was to drill in areas where the salt had been eroded or in regions whose characteristics were not conductive to oil and gas accumulations.[58] Identifying these spots required high-resolution geophysical profiles. In France, only the IFP, with logistical support from CNEXO, could provide them.

On September 29, 1973, the DSDP Executive Panel approved a second cruise across the Mediterranean and, two months later, they established the DSDP Mediterranean Advisory Panel to plan it. Second-time mission organizers Kenneth Hsü and Bill Ryan gathered a team of five experts from institutions across Europe, the Soviet Union, and the United States.[59] Marine geophysicist Lucien Montadert represented France. He was the only participant affiliated with an industry research institute—the IFP—rather than an academic research center. Montadert became the bridge between France's oil exploration sector and the international scientific community of the DSDP. His expertise in Mediterranean exploration, developed through numerous academic-industrial campaigns and advisory roles to oil companies, positioned him as one of the leading experts in the region. He was also deeply engaged in the scientific debate around the Messinian Salinity Crisis and a frequent participant in academic conferences like the CIESM.[60] Reflecting on his early career, Montadert later described the discussions held in the Mediterranean Advisory Panel as formative and invaluable for his professional development.[61]

The inclusion of Montadert was particularly strategic due to his access to multichannel seismic profiles, essential to identifying drill sites in presalt formations. Up until this point, the geophysical campaigns described in this book were single-channel surveys, which relied on a single sensor towed behind a ship. These systems collected data from a single point—and the sum of those points made up a seismic profile. In contrast, multichannel systems used an array of sensors working simultaneously, collecting data from multiple points along the seafloor. This resulted in greater penetration and higher resolution images, therefore enabling more accurate interpretations of complex geological structures. Yet, as happened with other technologies for marine geosciences, their use was initially limited to the oil industry due to their cost of operation and complex logistics. The IFP had pioneered the use of multichannel surveys in the western Mediterranean in the early 1970s to understand and date the deposition of the deep salt.[62]

Having Montadert on the Mediterranean Advisory Panel facilitated the exchange of these datasets and expert advice. He provided the mission organizers with the unique multichannel profiles collected by France's oil industry consortium (CEPM), alongside the geological and geophysical data that CNEXO and France's oil industry had jointly collected just after the first DSDP Mediterranean cruise. In other words, France's offshore network had invested money, manpower, and resources in producing new geological data and interpretations of the deep Mediterranean basin, which it now shared for an international oceanographic expedition. This information was instrumental in identifying drill sites that met both scientific and (presumably) commercial interests. Not only that—to address safety concerns, the IFP conducted a preliminary campaign to ensure that the drill sites it proposed complied with the DSDP Safety Panel's requirements.[63] Eventually, four out of seven drill sites approved for the *Glomar Challenger's* cruise were based on the IFP's data, half in the Balearic basin and half in the Levantine basin (in the Mediterranean's western and eastern sides, respectively).[64]

Because of his extensive scientific expertise and critical contributions to planning it, Montadert was appointed co–chief scientist of the second *Glomar Challenger* cruise across the Mediterranean, alongside Hsü. Joining Montadert on the scientific party was his IFP colleague Germaine Bizon, expert in Mediterranean microfossils (significant for developing both academic research and commercial assessment).[65] The *Glomar Challenger* returned to the Mediterranean on April 4, 1975, touring the basin for seven weeks, drilling into the abyssal plain, and recovering hundreds of meters of rock and sediment. After the expedition, core samples were distributed for postcruise studies across American, French, Italian, German, Soviet, and Swiss academic laboratories, as well as among the industrial laboratories of the IFP and French oil companies.

How should we interpret the apparent readiness of France's oil industry to contribute unique seismic profiles, human labor, and research resources to an international scientific campaign whose results would be openly published and shared? The short answer I propose is that, as scientific ocean drilling aimed deeper, so did the offshore oil industry. Between 1972 and 1975, during the planning of the second DSDP cruise, France's oil industry viewed participation in the DSDP as a cost-effective way to access deep-sea drilling technologies. This growing interest was likely spurred by the 1973 oil crisis, which shifted industrial focus toward exploring areas beyond the continental shelf.

The crisis began in October 1973, when the Organization of Petroleum Exporting Countries (OPEC) imposed an oil embargo on nations supporting Israel during the Yom Kippur War.[66] In six months, oil prices soared from 3 to 12 US dollars per barrel. This price spike heavily impacted Western economies. For France, the oil shock ended the *Trente Glorieuses*, the postwar period characterized by steady economic growth—now, the national economy stagnated while unemployment escalated. For the US, which imported 12 percent of its national oil consumption from OPEC countries, it triggered a frenetic search to secure alternative energy sources safer than the Middle East supplies.[67] As for the oceans, the crisis prompted the oil industry's attention to deeper regions. At 12 US dollars per barrel, it became economically viable to invest in technological innovation to install rigs beyond 300 meters deep. Previously unprofitable oil deposits now became commercially viable. The North Sea saw a boom in exploration in deeper waters, while in the Gulf of Mexico, the US government opened leases for drilling beyond 200 meters deep.[68]

But that commercial exploration on the abyssal plain became feasible did not mean that it was technologically effortless. Only the most advanced drillships could provide samples at depths beyond a dozen meters, and the scientific D/V *Glomar Challenger* was among these few. In France, the CFP had just acquired the D/V *Pélican*, a dynamically positioned drillship capable of drilling up to 320 meters deep.[69] However, its high cost of operation hindered its utilization as a reconnaissance vessel (that is, for surveys to gather preliminary data on poorly explored areas). Drilling a single deep-sea borehole cost about 10 million US dollars, and an entire exploration venture, including preliminary geophysical campaigns, postvoyage analysis of data, and the acquisition of exploration leases demanded about 5 billion US dollars. Because of the great expense of operation and burdensome logistics, after the *Pélican* came into use, the CFP prioritized drilling operations in areas already verified as promising, such as in the Gulf of Lion's continental shelf or in the shallow North Sea.[70]

From the perspective of France's oil industry network, thus, active involvement in the second DSDP Mediterranean cruise offered substantial benefits.[71] Sharing multichannel seismic data or conducting presite surveys using its own resources required little investment compared to the valuable returns: samples and data from deep areas that were not prioritized by their exploration drilling cruises.[72] In essence, France's oil industry participated in scientific ocean drilling for the same reasons the international scientific community organized cooperative ventures: to maximize access to cutting-edge technologies (specifically, a deep-sea drilling vessel) and enhance its own exploration capabilities.

France's oil industry relied on the information collected during the DSDP expeditions, as well as postcruise analyses from their laboratories on micropaleontology, geochemistry, and mineralogy, to assess the oil potential of the western Mediterranean—which fell short of initial expectations. In 1979, under CEPM's sponsorship, a team of petroleum experts from the IFP and the French oil companies CFP and Elf-Aquitaine drafted the report "Petroleum possibilities of marine regions in the western Mediterranean," to be distributed among the French oil industry network. According to the authors, none of the *Glomar Challenger*'s drill sites during Leg 13 and Leg 42 had showed hints of hydrocarbon deposits, although geochemical analysis had demonstrated that the Mediterranean basin's conditions favored the maturation of organic matter. Except for the deep Gulf of Lion, the western Mediterranean abyssal plain was still a poorly understood region, and potential hydrocarbon deposits would still require large economic and technological efforts to identify and exploit.[73] American and Spanish oil firms (like Chevron, Shell, and the Spanish ENIEPSA) had found valuable reservoirs only off the Spanish continental shelf.[74] It was still believed that massive hydrocarbon deposits were trapped under the massive salt layer, but French oil companies realized that they would not in the short term possess the technological capabilities to operate beyond 150 meters of depth.[75]

After the report was published, the oil industry's interest in the region began to fade. This was particularly evident when, in the framework of the second oil crisis beginning in 1979, France's oil industry redefined its exploration and production strategy in the offshore far from the deep western Mediterranean. The oil shock prompted, in less than a year, the price per barrel to rocket from 12 US dollars to almost 40 US dollars, stimulating oil companies to set commercial aspirations for offshore regions that were difficult to access.[76] The setting for offshore production changed globally. The Gulf of Mexico was lagging as a productive area: despite soaring oil prices, offshore leases were becoming prohibitively expensive (one could cost more than 2 billion US dollars) and discoveries too rare—although the region would experience a second deep-water boom in the 1990s and 2000s.[77] Meanwhile international interest turned to the North Sea as a new El Dorado. France's oil companies focused their efforts in already productive regions, namely the North Sea, where CFP and Elf participated in exploiting the oil and gas fields Ekofisk, Frigg, and Alwin North. Other productive regions included offshore Indonesia, the Gulf of Guinea, offshore Tierra del Fuego (Argentina), and the Atlantic's Iroise Sea. The extent to which this shift of commercial priorities caused the *Glomar Challenger* to not return to the Mediterranean is a question still to be answered, but the scientific drillship would not enter the Strait of Gibraltar to survey the entire basin again until 1995.[78] Although deep-sea drilling activities were not conducted for academic purposes for twenty years, the two DSDP campaigns across the Mediterranean left an important legacy for academic geology.

In 2018, almost fifty years after the *Glomar Challenger* offered evidence that a thick layer of salt covered the deep Mediterranean basin, the European project Saltgiant ETN kicked off, gathering scientists from European regions to resume investigating the Messinian Salinity Crisis. Many of its experts had studied under the mentorship of Montadert, Cita, or Ryan, demonstrating the creation of a scientific genealogy around the imaginary of a desertic Mediterranean. Collaborations with industrial partners were common, given the large number of offshore datasets that oil companies now possessed—with the high-quality, detailed resolution unimaginable at academic laboratories. Many of the events that took place during the Messinian Salinity Crisis, such as how the Mediterranean Sea refilled, whether the sea evaporated completely or only partially, or how climatic changes could have influenced it, are still to be thoroughly answered. Nevertheless, the deep Mediterranean seafloor is no longer an imagined space: it can be *seen* through novel three-dimensional seismic profiles, experienced through deep-sea cores, and understood through international cooperation.

DRILLING THE SEAFLOOR: INTERSECTING
NARRATIVES IN FRANCE

Scientific deep-sea drilling cruises were not designed to pursue commercial goals, yet their findings often intersected with industrial needs. In a context where geological results could advance knowledge of the oceans' history while also providing insights into hydrocarbon deposits, the relationship between scientific research and industrial interests cannot be overlooked. Drillships are exclusive and groundbreaking technologies, both extremely costly and logistically complex to operate. Scientists from various regions sought to use them to access the deepest parts of the seafloor, and industry stakeholders had similar aspirations.

During the 1970s, the French oil industry became increasingly involved in the DSDP campaigns in the Mediterranean, leveraging the results to reduce the time and cost of deep-water exploration. While France's oil firms supported the exploration of the continental shelf, the deep seafloor presented a far more daunting and expensive challenge. Obtaining deep-sea cores using a private drillship was prohibitively expensive, but by relying on the *Glomar Challenger*'s results, oil companies were able to access critical geological information while reducing the time and cost invested. For French oil industry managers, participating in these campaigns—whether through data sharing, site selection, or post-cruise analyses—was a strategic decision that provided direct access to valuable information. In this sense, the narrative of a scientific project and a unique geological discovery is inseparably linked to the history of France's offshore industry and vice versa. We cannot fully understand how we came to know the deepest Mediterranean seafloor, its salt domes, and its compelling past, without recognizing the active role of France's oil industry.

Given these entangled scientific and industrial stories, we should revisit Hsü's narrative on the discovery of the Mediterranean salt giant (as this massive deposit came to be known among geoscientists), to discuss what this case reveals about the marine science-industry relationship. There are three main reasons why he might have decided to publish the book: to preserve the memory of a revolutionary discovery; to popularize marine geosciences and scientific ocean drilling among lay audiences; or to strengthen his position in the ongoing scientific debate—most likely, a combination of all three. However, between the lines, we can glimpse his attempt to present scientific ocean drilling as a purely scientific endeavor, in which industry stakeholders or commercial interests played no role. This position is understandable since, by the mid-1970s, rising environmental concerns began to cast a shadow over offshore activities. With broader public opinion condemning the oil industry's operations in the oceans, scientists felt compelled to emphasize the distinction between their work and hydrocarbon prospecting. Yet, as this chapter has shown, the contributions of the oil industry were undeniable: scientific reports acknowledged industry stakeholders for their data and surveys, young geologists and geophysicists connected to offshore industries were heavily involved in the planning and development of the expeditions, and they became deeply integrated in the scientific community.

In this sense, this book closes with a case where the paths of international science and France's offshore industrial development intersected, set against the backdrop of the Mediterranean. However, there are other cases still to be explored that transcend these geographic boundaries. I will reiterate this: scientific ocean drilling is (and has been) a scientific endeavor, but it wouldn't exist as we know it without the contributions of offshore industries—and not just in terms of technologies and drilling vessels. The expertise of its workforce, from roughnecks to engineers and technicians; safety standards; the logistics of a cruise; offshore operations; and shore-based management—all have shaped scientific ocean drilling from its inception, most acutely in the first decades. Acknowledging the essential interplay between science and industry in seafloor exploration does not diminish the environmental concerns or controversies associated with the oil industry. Rather, it is a call for a nuanced understanding of its role, focusing on the specific ways in which it has contributed to our current capabilities for seafloor exploration.

Epilogue

And now, how can I capture the impressions that this underwater
stroll has left on me?
Words fall short to describe such wonders!
JULES VERNE, *TWENTY THOUSAND LEAGUES UNDER THE SEA*

A FADING VISION

In January 1982, an article published in *Le Monde* announced that, in official
discourses about *oceanology*, "realism has replaced lyrical illusion." Jour-
nalist Yvonne Rebeyrol—who had been leading the mass media coverage of
CNEXO and international ocean affairs in France—observed that, for the
last fifteen years, policymakers had overused and exhausted the mantra of
"the oceans as key to the world's future." Now, results were revealing the
poor relation between those promises and the economic performance of
actual ocean exploitation: in France, no offshore hydrocarbon deposit had
been discovered within its marine territory, while fisheries and their byproducts operated at a trade deficit, and marine pollution remained poorly
managed, regulated, and prevented.[1]

This headline felt like it shattered a spell, as if France was awakening from
a long dream in which public officers, industrial experts, and the wider population had been immersed. For almost two decades, the oceans and their
seafloor had occupied a prominent position in the collective imaginary as
spaces over which new industries, the national economy, and France's international prestige could be built. Ocean exploration made headlines in newspapers, which celebrated with fanfare each new geological or economic
discovery, as if those were sweeping France toward that promising future.
The conception of the seafloor changed into a space where strategic alliances could be established and across which a new territorial policy could
be performed. Official documents, public discourses, and mass media depicted the seafloor and its resources as the basis of a powerful, autonomous,

and independent France. But paradoxically, the same research fields, institutions, and technologies—developed to reify that future—in fact revealed its real limits and vulnerability—because the deeper we know a territory, the more realistic become the projections and predictions we can make. Marine geological studies and geophysical surveys showed that the French Mediterranean continental shelf did not harbor hydrocarbon deposits large enough to sustain the hexagon's supplies. Limitations on industrial technologies would hinder the extraction of oil and gas from thousands of meters' depth for decades to come. After years investigating the economic potential of polymetallic nodules beneath the Pacific Ocean, it became evident that relying on their extraction for industrial supply was unfeasible.

This reality check was accompanied by the progressive introduction, in the popular and political consciousness, of the oceans as fragile, finite, and unknown environments. In France, between the sixties and the eighties, a number of oil spills devastated regions along the northern and western coastlines of the hexagon, alerting the population to the harmful consequences of oil at sea. After the 1967 disaster of the oil tanker *Torrey Canyon*, in January 1976 the *Olympic Bravery* split in half on the coastline of Ushant, a French island twenty kilometers off Brittany, pouring into the sea more than a thousand tons of crude oil. Two years later, the crude carrier *Amoco Cadiz* ran aground on Portsall Rocks, two kilometers from the same coastline, producing the largest oil spill to date: more than 200,000 metric tons of oil spread through the ocean waters and along the coastlines of France.[2] The disaster also spread the population's rage against French authorities for their incapacity to effect the cleanup—especially after *Le Monde* disclosed that, two months before the disaster, an official internal report had denounced France's inadequate means to respond to offshore oil spills.[3] In the collective imaginary, the devastating harms of transporting crude oil were also directly attributed to the exploration and production of offshore hydrocarbons, accelerating the populations' rejection of any industrial activity over the seafloor. As environmentalism gained political popularity, so did the oceans. The first "green" political party in France, the Greens-Ecologist Party (which ran a candidate in the 1974 presidential election), introduced an environmental consciousness among the French population that included the protection of the oceans from pollution and biodiversity loss. At the same time, around the world numerous ecological movements exerted pressure to regulate the oceans' conservation. In 1982, the United Nations Convention on the Law of the Sea was approved. Besides establishing national and international regulations over the ocean territory, it instituted a regime for protecting and preserving the marine environment, regulated the exploitation of mining and fishing resources, and prohibited the dumping of radioactive waste in the oceans.

Captain Jacques Cousteau, archetype of the passion for the oceans, reflected this shifting attitude toward the marine environment. As shown throughout this book, Cousteau began his media career by disseminating his fascination about the oceans' infinite potential for future modes of human life: food, energy supplies, or space for settlement were only some of the resources that humans could grab from the oceans. By means of technoscientific development, humans would be able to tame and tailor this new territory full of promises. However, Cousteau's position toward the oceans progressively changed, hand in hand with coeval environmentalist ideas and accumulated knowledge of the marine environment. In the early eighties, Cousteau grew fully conscious of the power of his influence.[4] His feature films, projected in more than a 120 countries around the world, began to be specifically designed to convey a more powerful message beyond showing the marvels of the underwater: that the oceans were spaces vulnerable to human activities. Technoscientific development was no longer a means of "conquest" but should be aimed at harmonizing ecology and economy. His renewed message, aligned with simultaneous discussions taking place at the United Nations and within environmentalist organizations, advocated for preserving the oceans for future generations.

THE IMPACT OF THE (OCEANIC) SCIENCE-INDUSTRY STRATEGY

Perhaps to reinforce his new stance, Cousteau published in 1981 the booklet *French People, We Stole Your Sea* (*Français, on a volé ta mer*), coauthored with his financial advisor and long-time colleague Henri Jacquier. In it, Cousteau and Jacquier harshly attacked CNEXO, the French administration, and their expansionist aspirations over the oceans during the previous two decades. In their view, the ambitions of a political elite, driven by delusions of *grandeur* and unmeasurable richness, had provoked unjustifiable, massive expenditures in ocean sciences (almost 400 million US dollars overall) that paradoxically, as they argued, had barely had any positive impact on France's economy and marine research.

In economic terms, although CNEXO had been designed to cover its expenditures through the extraction of ocean resources, its dependency on the public budget never decreased from 85 percent.[5] Quoting Cousteau's words, the cause of this failure was found in the frantic (political) quest for resources in the oceans, a situation he acidly depicted this way:

Marine spaces triggered the imagination of journalists as well as the appetite of accountants. The unit of reference was the million tons: million tons of oil, of animal protein, and of metals were to fulfill our hunger for energy, food, and raw materials. The nourishing sea became a large department store, the cave of Ali Baba, where a prodigious inventory paraded before our amazed eyes. [...] The images evoked create a sensation of fairy tale. These riches "sleep" under the sea. Only the kiss of a charming prince would awake them. The State, which loves to be disguised, was tempted by the role.[6]

Cousteau and Jacquier argued that oceanography became a tool too powerful to leave in the hands of oceanographers and marine geoscientists, as the third chapter here discussed. In a period when the oceans and their seafloor were overtly regarded as a new El Dorado, public officers and policymakers came to consider themselves, in Cousteau's words, the most suited to judge "the supreme interests of the oceans for the nation."[7]

About the impact of CNEXO's policies on marine sciences, Cousteau and Jacquier outlined that only those expert teams who had aligned their research interests with the national, centralized agenda had benefited from CNEXO's budget and activities. Meanwhile, researchers at public universities continued to suffer an acute shortage of research means. This criticism is not unique to the case of France's industrial patronage of ocean exploration, but it emerges as a recurring theme in such state-led patronage after World War II. As the historian of science Naomi Oreskes has argued, during the Cold War the US Navy provided substantial resources to research teams and laboratories whose interests aligned with the topics the navy needed to know (research on currents, tides, ocean bottom sediments, seabed topography, etc.). In contrast, researchers devoted to fields that did not directly address those needs, like marine biology or chemistry, struggled to find generous patrons to advance their research interests.[8] The Cousteau Group felt into the latter category, originating a resentment that added to Cousteau's strained relationship with General Director Yves la Prairie. In a paper recalling these years, Jacquier recounted that, since CNEXO's funding, Cousteau had felt neglected by CNEXO's administrators in the field of ocean exploration, even though he had dedicated his career to it. He sensed that his words were not listened to in CNEXO's managerial circles, that his role at the Scientific and Technical Committee was sidetracked, and—perhaps more formative to his harsh criticism—that he failed to secure stable CNEXO funding for his marine expeditions.[9] Cousteau's ideas about the oceans, the type of exploration that ought to be pursued, and the technologies that France should develop did not converge with the CNEXO agenda. This did not mean that Cousteau opposed industrial support for ocean exploration. Quite the contrary, until the late seventies he eagerly collaborated with offshore industries, for example, to design submersibles to repair offshore platforms or study human physiology at great pressures, which would eventually allow experts to live underwater at industrial infrastructures. These projects had provided Cousteau with funding, experience, and underwater technologies to advance his main ambition: to foster a deeper understanding and appreciation of the underwater environment. This basic principle simply did not fit CNEXO's bill—a national interest to explore "the riches of the seas."[10]

Apart from his personal standpoint and relation to CNEXO, Cousteau's assessment is relevant because it conveys the value the oceans acquired in a context in which geopolitical frictions, future uncertainties in a restructured world, and the material basis to sustain an ever-growing world population pervaded political discussions and public discourses. Economic futures anticipated by a political elite constituted a window of opportunity for particular groups of marine geoscientists in their pursuit to explore the seafloor. In a compressed time period, a series of political, economic, and industrial anxieties and interests converged in the need to know an unknown region, prompting unprecedented public and private investment in developing marine geophysical techniques and geological knowledge. With CNEXO, ocean sciences were reserved to promote state interests. The seafloor became an unknown gameboard for the new world order, while marine geosciences were used to establish diplomatic relations, explore new sources of energy and material resources, and strengthen France's position in issues of territorial legislation. But, as Cousteau pointed out, the government's investment in ocean exploration did not benefit equally all communities of French geoscientists. This book has focused on a number of academic and industrial communities that collaborated in the national oceanic program, leaving aside multiple groups and laboratories whose research largely depended upon the budget available at France's National Research Council. Despite the uneven patronage, the experts who witnessed firsthand the birth of French marine geosciences, whether from university laboratories or industrial institutions, agreed that the sixties and seventies were a glorious period for seafloor exploration.[11]

THE CONSTRUCTION OF NEW DEEP TERRITORIES
(AND THE MECHANISMS TO ACHIEVE IT)

This book has sought to historicize the seafloor, by showing how it progressively became part of our worldview as it was transformed into a three-dimensional territory—that is to say, a space conductive for human activities that was explored, tamed, and known through new technological practices. Economic imaginaries of future resource supplies fueled its exploration, while research fields, technologies, organizing structures, bodies of experts, and new multidimensional collaborations coevolved hand in hand with the seafloor's shape, depth, and detail.

The case of France offers a revealing example of the interplays between the construction of submerged territories and the state's priorities. The concerns of a declining colonial power, immersed in political crisis, led its public officials to project the oceans as the pillars of their country's future economic security and international power. For some industrial managers and public officers, the seafloor came to be regarded as "the replacement of the Sahara," envisaging there equivalent economic motivations and developing similar mechanisms of exploration and exploitation as in former and lost French colonies.[12] The seafloor took its shape as a three-dimensional territory at the same time that its definition as a new economic frontier expanded—and, with it, the hopes of the French administration for the promise of this new space. This evolution is represented in the place that ocean exploration occupied in France's political agenda. In 1958, exploring the seafloor was considered one more component of the ocean sciences, embedded in a rhetoric of promoting "fundamental research" to increase the nation's prestige. A decade later, it had transformed into a central asset in the country's industrial strategy, an avenue to recover its international prestige and exercise its diplomatic relations.

The political intention alone was not sufficient, however, to render the seafloor a new territory to explore. The key to France's success in ocean exploration lay in the alignment of interests of disparate expert groups, a convergence that was not a serendipitous coincidence. Bringing together dispersed and disconnected communities required effort, negotiations, and a well-planned strategy that became invisible as it drove scientific research. First, land-based geologists gained a foothold in the national oceanographic agenda by deliberately outlining the scientific elements more pertinent to the national program's development. Their research plunged into the Mediterranean Sea (metaphorically and literally) when geologists like Jacques Bourcart allied with naval officers and oceanographers. For the first group, it was Bourcart who accommodated his research to the given needs of submarine defense while, for the second, alignment pivoted on a new fraternal sense of sharing an (underfunded) working environment. In other words, at the same time the seafloor transformed into a research environment for geologists, it came into the perception of other communities, which in turn also integrated it into their operational space. Yet it was CNEXO that steered the growth of a network progressively connecting university researchers, petroleum geologists, industrial managers, and public officers, driving them in a common direction.

CNEXO offers an illustrative case of a well-planned strategy to align disparate interests, which became invisible when driving scientific research. The institution functioned as the key mechanism for articulating scientific production with commercial exploitation and national goals. It was planned by influential stakeholders interested in different aspects of the oceans but who were not familiar with the specifics of ocean research. Public administrators, entrepreneurs, representatives of offshore industries, and bureaucrats wanted to obtain detailed information about the seafloor, the mass of water, and its living beings to foster their marine ventures and aspirations. They gathered to seek insights into the marine environment, its potential resources, and its geopolitical usefulness with little concern for the scientific means to achieve it. In other words, CNEXO functionality resided in the disconnection between how leading circles proposed goals and produced guidelines, and how those became embedded in marine cruises, scientific programs, and laboratory research. The CNEXO administrative board supported what they called "basic research" because, in their view, it was the only way to engage scientists in their plans, but also because researchers were the best prepared to decide how to explore the oceans.

I have called "invisible" the strategy that drove research to the extent that scientists were not overtly compelled, through their research, to identify hydrocarbon deposits, estimate the reserves of sands and gravels on the continental shelf, or assess the economic potential of polymetallic nodules in the deep ocean bed. Researchers freely designed their investigations to fit within the agenda requirements. They had the autonomy to select their research goals, design marine campaigns, settle scientific collaborations, and choose the methods for collecting, processing, and interpreting data; strong secrecy policies did not bind the publication of data or research results. That's why, when we look at the scientific and technological outcomes of CNEXO, we don't detect an explicit intent of resource prospecting or production, and that is probably the reason why the institution's initial commercial vocation has faded into oblivion. This argument about researchers' intellectual autonomy in patronage relations is far from new, especially in the recent history of the oceans. For the last twenty years, scholars like Naomi Oreskes, Jacob Hamblin, Ronald Rainger, and Ronald Doel (to name a few influential examples) have discussed in great detail the meaning of epistemological freedom and research autonomy in the context of military patronage of the ocean sciences.[13] But I consider it pertinent to revisit the argument in this new framework—industry sponsorship and resource exploration, not directly driven by national security concerns—because it evidences that it is not a particularity of military contexts, nor is confidentiality a major shaping force. The commercial-scientific partnership drove the growth of marine geosciences in France in a way similar to how the military-scientific alliance steered the flourishing of oceanography in the US. In both cases, the ability to design a unified structure with common goals creates—borrowing the term Naomi Oreskes coined—a "context of motivation" for scientists to join.[14] CNEXO formalized pathways of industrial-academic collaboration and established carefully selected diplomatic alliances to explore the oceans and build France's new submerged territory. For marine scientists, this meant not only a rich source of funding but also a hub of research opportunities, promising paths of career development, and new partnerships at home and abroad. It is important to note that this setting was facilitated by the idiosyncrasies of France's oil industry which, instead of being a coalition of private firms, was composed of institutions attached by design to governmental structures. This distinctive arrangement not only enabled a smooth articulation of priorities from political cabinets regarding activities related to extraction, but it also promoted the design of a national, coordinated strategy that integrated academic research closely with industrial activities. Whether CNEXO reflects a unique attempt at national coordination to explore the oceans or a widespread, transdisciplinary strategy also implemented in other countries, remains to be determined by future academic studies.

The seafloor's territory was not only a rhetorical component in political discourses, but a material reality opened up via technological innovation that expanded national frontiers steadily into deeper areas. This book has followed the development of the seafloor's construction, from the surface of the shallow Mediterranean continental shelf, toward a three-dimensional perspective of the seafloor in the deepest oceans, and culminating in the acquisition of the capability to experience the seafloor directly, through deep-sea cores and with manned submersibles. In this process, the notion of "deep" transformed. For geologist Jacques Bourcart, "deep" meant surface sediments from near the shore. For the incipient offshore oil industry in the fifties, "deep" was considered any region beyond a dozen meters of water. In the 1970s, "deep" became the drilling limit of the *Glomar Challenger*, which could recover rock samples thousands of meters below water and seabed, triggering the oil industry's interest in those regions; and for the mining industry, "deep" was the location of polymetallic nodules in the abyssal plain. By employing the phrase "new deep territories" in the title of this book without explicit reference to the marine environment (by perhaps adding the terms "submarine" or "underwater" to "territories"), I aimed to evoke the notion that the seafloor came to be conceived as an intrinsic part of the national territory, a piece of land as conducive to human activities as the mainland—despite the essential technological mediation. Describing the seafloor territory as "new" points to the fact that, before developing the research means that enabled its exploration, it was completely beyond human perception.

But the anticipated oceanic economy never arrived; nevertheless, France's quest to control ocean resources had long-lasting consequences. It enabled the birth of marine geosciences, led to new understandings about the seafloor, and stimulated the establishment of cooperative relations between academic geologists and oil experts that have persisted to the present.

FROM PAST TO PRESENT: CONTINUITIES AND DISCONTINUITIES IN THE SEAFLOOR'S EXPLORATION

CNEXO's adventure ended in 1984, soon after the chronological ending point of this book. After inaugurating a Ministry of the Sea that would embrace all kinds of national activities related to the oceans, the office of President François Mitterrand decided to fuse the two big national institutions devoted to the oceans' exploration and exploitation—CNEXO and the Scientific and Technical Institute of Maritime Fisheries—as the French Research Institute for Exploitation of the Sea (IFREMER).[15] The new center was designed to focus on the management of pollution and on research resources (like submersibles or oceanographic vessels), leaving aside the industrial production of marine resources. Today the IFREMER sits at the Oceanological Center of Brittany, serving as the managerial center of France's oceanographic fleet and a renowned research center.

If this story was to finish here, we would be left with a linear narrative in which an imaginary of conquering submerged riches was replaced by a more realistic overview of the seafloor's resources and grounded evidence of the environment's fragility. However, such a narrative would obscure the three aspects that still cohabit (and interact with each other) in the exploration of the seafloor: a fascinating imaginary of treasures to be discovered that powers technoscientific development, geopolitical frictions derived from the use and control of the submerged territory, and a new realization of its fragility and finite nature. In other words, even though we are now more aware than ever of our limited knowledge about the deep ocean and its seafloor, and about the irreversible damages caused by human activities there, the seafloor's exploration continues to incite visions of new pathways of economic progress and resource autonomy. Technoscientific development is still deemed the solution to harmonize both approaches.

It continues to have the potential to open new exploitation thresholds and, therefore, to generate new geopolitical frictions. The quest for offshore hydrocarbon deposits continues as the oil industry's frontier marches onward into deeper waters. In the Gulf of Mexico, the platform *Perdido* by Shell Oil Company produces crude oil from almost 2,500 meters deep. Such a technological achievement conveys the message that offshore oil exploitation is now possible beyond the continental shelf. This leads to new questions about controlling deposits beyond national waters, but it also prompts considerations about the kind of energy sources that should sustain future human life. The ocean is at the center of international concerns regarding the impact of human activities on the environment. Numerous global summits and intense scientific research are aimed at addressing issues like sea-level rise, water warming, and depletion of biodiversity; spills of oil, chemicals, and industrial byproducts; the pervasive presence of microplastics drifting across the entire water column; and trash that pollutes even the most remote corners of the ocean floor.[16] At the same time, these forums seek solutions to sustain a human population in constant growth, leveraging the oceans and their seabed. The current challenge is how to reconcile these two aspects of the relationship between humans and the oceans.

Moving to shallower waters, the Mediterranean Sea has become a theater of conflicting interests and opposing management strategies. In the western basin, France, Spain, and Italy are progressively halting offshore prospecting and production of hydrocarbons through various legal mechanisms, moved by environmental concerns and the growing pressure of public opinion. Meanwhile, in the eastern basin, recent discoveries of massive gas deposits spread across the national waters of Israel, Egypt, Cyprus, and Turkey have strained geopolitical tensions. To the inherent difficulty of agreeing over the limit of maritime frontiers are now added unilateral proclamations to expand them, jeopardizing the political stability of the region in the years to come.

Potential mineral resources hidden beneath the high seas are also in the spotlight of national interests and international negotiations. In June 2022, France's President Emmanuel Macron reintroduced the oceans to the core of the national political agenda for geopolitical and economic reasons. The French Senate announced a five-year program, with a budget of 3 billion euros, to develop the technoscientific capabilities that will enable France to explore (and eventually exploit) the deep ocean floor beyond the threshold of a thousand meters' depth. The imaginary of the deep ocean as a space full of riches still to be discovered, whose rational exploitation could be balanced with its conservation, is explicitly articulated in the new national program: the ocean floor appears, once again, described as a new

El Dorado, a region to be "conquered," "the last frontier" to be explored—a discourse that strongly resembles the one that fueled ocean exploration six decades earlier.[17] In this framework, scientific and technological development are of central importance. New technologies and knowledge are deemed the keys to open untapped oceanic riches, and oceanographic cooperation becomes a means to maintain cordial diplomatic relations in exploring vast spans of international waters. Only the future will tell which parts of the story here presented will be replicated, and where our present understanding of the marine environment might spark a divergence.

Between the past and the present there are also discontinuities that we should take into account to better understand how marine exploration has changed. Among the notable characteristics of the time period considered in this book were the alliances, patronage relations, and collaborations between industrial stakeholders and academic experts. In the early years of seafloor exploration, these relations benefited both communities because the technological and skill gaps that separated them were relatively small. Whereas the offshore oil industry possessed larger budgets and more advanced research resources, university geologists and geophysicists contributed valuable scientific expertise. The seafloor was an unknown for both groups when they converged on the priority of investigating its characteristics in a most pragmatic way. However, by the early eighties, the offshore oil industry became increasingly specialized, developing technologies specifically suited for hydrocarbon prospecting, and starting to rely on new, private consulting companies rather than on academic scientists to interpret data. Nowadays, the oil industry possesses data of a quality unattainable for academic researchers—notably three-dimensional seismic profiles, which have opened a new pathway for perceiving and understanding the seafloor's structure. Only occasional academic-industrial collaborations are established, and the exchange of data is not as frequent as in the sixties and seventies. Both communities follow different pathways, even though the regions explored and the geological interests at stake might, in theory, converge.

Similarly, scientific ocean drilling programs still exist and thrive in a linear genealogy, yet the involvement of the oil industry in them is almost nonexistent for the same reason—namely that the oil industry possesses more advanced drilling and research technologies than those available for scientific research. Starting in the mid-eighties, the goals of scientific ocean drilling have turned decisively toward creating knowledge about climate change, geohazards, and different modes of life that are still to be discovered deep beneath the seafloor by understanding the ocean crust. Leadership has also changed. Formerly an entirely American program, leadership and decision-making in ocean drilling have been progressively distributed among other partner nations and organizations. More than twenty countries have participated within different regimes and consortia, such as the European Consortium for Ocean Research Drilling (ECORD), the Australia and New Zealand Scientific Drilling Consortium (ANZIC), Japan, India, Brazil, Canada, South Korea, the People's Republic of China, and even the former Soviet Union and (for a short time) Russia. At the moment I write these lines, scientific ocean drilling is undergoing the biggest transformation since its inception in the 1960s. In January 2025, Japan and ECORD joined forces to lead the program's new phase, the International Ocean Drilling Programme (IODP-3, as the third iteration using this acronym), which other nations will also join.[18] For the first time, the US has paused its participation in the international program, coinciding with the retirement of the only US scientific drillship, the *JOIDES Resolution*, after more than four decades of service to international geosciences—much to the sorrow of the global ocean drilling community. While geoscientists worldwide have urged the NSF to continue its participation, the future shape of US involvement—whether through a new national program or a newly dedicated scientific drillship—is still under discussion. At the same time, China has just launched its scientific drilling vessel, the *Meng Xiang*, which will serve as the centerpiece of its national scientific ocean drilling program. This development paves the way for unprecedented collaborations across different ocean drilling programs. The consequences of these transformations, their impact on scientific research and on international cooperation, will undoubtedly unfold in the coming decade.

The seafloor is a contested space that mirrors national concerns and international relations, and this reality is as important for understanding the past as it is for understanding the present. Since World War II, it has been a gameboard for extractive imaginaries, power balances, geopolitical strategies, and emergent industries. It was a catalyst for the reconfiguration of power relations in the postcolonial world order, and it is now a barometer for understanding the future of international relations. At its center were, and still are, the promises of yet-to-be-discovered, and accessed, natural resources in the oceans.

When imaginaries dissolve, what remains is the physical space that has been created. Human activities, oceanographic missions, acquired data, unsuccessful industries, relationships established, and institutions can disappear, fall into oblivion, or fuse together and lose their identities. What remains, instead, is the territory as a space inside the limits of human perception, tailored to our use through technological devices. This applies to the Wild West, the Antarctic continent, or Earth's airspace as well as to the oceans and their seafloor.[19] It can also apply to future territories, still to be explored. Some of the ideas and conclusions here can be useful as a frame of reference to interpret the processes and consequences derived from the exploration of future unknown (and still imperceptible) regions such as, for instance, other planets in outer space. Imaginaries precede technological capabilities and real knowledge. Then new institutions, unexpected collaborations, and bodies of skilled experts can emerge, and different stakeholders can ally or confront each other to achieve control of such future territories. Regulations emerge always one step behind the imagined uses of the territory.

Acknowledgments

Finishing this book feels like the end of a journey that started with research in France, developed into a book manuscript in Spain, and is being completed during my time as a postdoc in Japan. Along the way, I have been fortunate to receive insights, guidance, and encouragement from numerous people and institutions. While it would be impossible to list everyone who contributed their time, expertise, or support, I want to express my deep gratitude to all those who have been part of this process.

Early work on this book benefited from funding provided by the Deepmed Project (ERC-CoG DEEPMED-101002330), under Lino Camprubí's direction. I am deeply grateful to Lino not only for his support, but also because his works inspired my interest in the history of the deep ocean from the very beginning.

Being immersed in the marine geosciences community has been an essential and enriching experience since the start. This book pays tribute to the geoscientists, administrators, and operators who strive to uncover the secrets of the seafloor and who have shared their knowledge with me. Countless conversations, from oral history interviews to informal chats, have shaped the story in these pages. Many of these people offered their time and shared their pasts with me—I am indebted to every interviewee. I hope this work (with its inherent limitations) does justice to their collective past. All mistakes are my own.

The background and some of the ideas were born during my time at Sorbonne University in Paris, as part of the interdisciplinary European project Saltgiant ETN. It was there that I first became fascinated by the seafloor and its exploration, the work of marine geoscientists, the challenges of at-sea operations, the complexity of funding routes, and the hospitality of the community. I am sincerely grateful to Néstor Herran for his patient and thoughtful mentorship, which helped me grow as a historian. Alongside him, Laurent Jolivet and David Aubin offered valuable assistance in both dimensions, science and history. The Saltgiant community, from senior researchers to my fellow early-career colleagues, kindly taught me how to look at the seafloor through their techniques, shared their joy in research, and showed me what it means to belong to the scientific community. I owe them all my heartfelt thanks.

The final stages of this book were completed in Japan, where I am currently a postdoctoral researcher at the Japan Agency for Marine-Earth Science and Technology (JAMSTEC). I am grateful to Nobu Eguchi and my team for enabling me to dedicate a portion of my time to finalizing the manuscript while continuing my research on the history of scientific ocean drilling. My involvement with the Institute for Marine-Earth Exploration and Engineering (MarE3) over the past two years has kept me closely connected to the scientific ocean drilling community. The countless conversations and shared experiences with them have enriched this book, adding nuance and perspective to its pages.

I feel immensely fortunate to have worked with Helen Rozwadowski and Katherine Anderson as series editors. Their enthusiasm for the project, their invaluable feedback, and their steady guidance at every stage have been indispensable. I cannot thank them enough for their dedication. Along with them, Karen Darling and Fabiola Enríquez at the University of Chicago Press have been exceptional editors, always ready to assist and guide me through the publishing process. I extend my appreciation to the two anonymous reviewers whose thoughtful feedback significantly improved the manuscript.

Our work as historians would not exist without the crucial aid and efficiency of librarians and archivists. I want to express my gratitude to those at the institutions that have provided me invaluable access to resources and collections.

Finally, my family and my husband's family have always shown their enthusiasm for and pride in my work, which has been a source of warmth and strength. My husband, Javi, has been my shipmate throughout this journey (and many others). He has been by my side in both calm waters and turbulent storms, helping me to hold the helm during times when I struggled to stay on course, always guiding me back on track, and consistently reminding me to keep the final destination in sight.

Notes

All quotations originally in French have been translated by the author.

INTRODUCTION

1. About the history of underwater cables, see Starosielski, *Undersea Network*.

2. An "imaginary," here, means not only an imagined, fictional, individually held representation, but a vision crafted and held by a large community about the seafloor's role and possibilities for human life. "Sociotechnical imaginaries," first coined in Jasanoff and Kim, "Containing the Atom," are defined as collective visions of good and attainable futures that constitute driving forces for technoscientific development and social construction (McNeil et al., "Conceptualizing Imaginaries"; Sismondo, "Sociotechnical Imaginaries").

3. Ocean scientists and social scientists have frequently inquired why wide audiences don't think about the deep sea. For a detailed study providing answers that range from perception to psychology and geography, see Jamieson et al., "Fear and Loathing."

4. About territories, see Elden, "Land, Terrain, Territory"; Peters et al., *Territory Beyond Terra*.

5. For in-depth analysis of the coconstruction of decolonization, techno-utopian imaginaries, and the ocean's postcolonial governance, see Ranganathan, "Decolonization and International Law"; Ranganathan, "Ocean Floor Grab"; Robinson, "Scientific Imaginaries and Science Diplomacy."

6. Miles, "Technology, Ocean Management, and the Law of the Sea"; Barkenbus, "Politics of Ocean Resource Exploitation."

7. Enumeration based on Mero, *Mineral Resources of the Sea*. Also expressed during benchmark speeches (UN General Assembly, "First Committee Debate," November 1, 1967).

8. Ranganathan, "Decolonization and International Law"; Ranganathan, "Ocean Floor Grab." After two decades of resource exploration, how much production was achieved, and to what extent did it remain a future techno-utopia? To understand the state of ocean mining and resource production in the eighties around the world, as well as the problems they opened internationally, see Mann Borgese, *Mines of Neptune*.

9. Kroll, *America's Ocean Wilderness*; Robinson, "Scientific Imaginaries and Science Diplomacy"; Turner, *Significance of the Frontier*. For a critical perspective on the American frontier, emphasizing its ideological and cultural role, see White et al., *Frontier in American Culture*; Rozwadowski, "Ocean's Depths"; Rozwadowski, "Arthur C. Clarke"; Rozwadowski, "Engineering, Imagination, and Industry." See also Rozwadowski, "'Bringing Humanity Full Circle."

10. See, e.g., Finley, *All the Boats on the Ocean.*

11. For the US SEALAB projects, see Squire, *Undersea Geopolitics,* and for the Canadian, Adler, "Deep Horizons." Cousteau's projects to inhabit underwater appear in Matsen, *Jacques Cousteau.*

12. About the history of the offshore oil industry, focused on American firms, see Priest, *Offshore Imperative* (on Shell's technological innovation in exploring the offshore); Pratt et al., *Offshore Pioneers,* on the oil-tech company Brown and Root; Burleson, *Deep Challenge!,* a narrative based on archival documents from the oil-tech company Glomar Inc. From the history of science and technology, see Van Keuren, "Breaking New Ground," about the oil industry origins of technologies for scientific deep-sea drilling.

13. The concept of "environing technologies" has recently been coined to stress how technologies construct the environment: see Sörlin and Wormbs, "Environing Technologies"; Gärdebo, "Environing Technology." For the mediating role of technologies and the marine environment, see, e.g., Rozwadowski and Van Keuren, *Machine in Neptune's Garden*; Höhler, "Depth Records and Ocean Volumes."

14. United Nations, *Convention on the Continental Shelf,* Article 1.

15. Robinson, "Scientific Imaginaries and Science Diplomacy." On how marine technologies open new thresholds of knowledge, see Rozwadowski and Van Keuren, *Machine in Neptune's Garden.*

16. Negotiations at the Law of the Sea conventions have been thoroughly analyzed from the perspective of legal history (Miles, *Global Ocean Politics*; Hollick and Osgood, *New Era of Ocean Politics*; Hollick, *U.S. Foreign Policy and the Law of the Sea*; Churchill and Lowe, *Law of the Sea*). About the coalition of seventy-seven nations (most of them from the Global South) who gather to promote their collective interests regarding access, governance, and management of ocean resources, see Friedman and Williams. "Group of 77."

17. Quéneudec, "La France et le droit de la mer."

18. Fernand Braudel argued that geography was the basis of political, economic, social, and cultural life; he focused on the Mediterranean basin to analyze these relations in the sixteenth century (Braudel, *La Méditerranée et le monde méditerranéen*). Beginning in the mid-nineteenth century, ocean sciences (especially marine biology) became political tools to rebuild France's power and international prestige. See Adler, "Marine Science for the Nation or for the World?," on how marine stations on the Atlantic coastline were designed to rival English advances in marine science and spread France's international prestige. Beyond France's case, the oceans have been pivotal areas for the development of empires. For a science, technology, and environmental perspective, see Reidy and Rozwadowski, "The Spaces In Between," and Smith, *To Master the Boundless Sea.*

19. During the fifteenth and sixteenth century, France's colonial empire included territorial possessions on the American continent (in today's Canada, the USA's Louisiana, the French Antilles in the Caribbean, and French Guiana). France also controlled a region in today's India. Known as the First French Colonial Empire, it collapsed in the early nineteenth century.

20. Hecht, *Radiance of France*; Pritchard, *Confluences.*

21. Camprubí and Lehmann, "Scales of Experience." On the interplay between history of science and technology and environmental history to study the oceans, see Rozwadowski, "Oceans: Fusing the History of Science and Technology with Environmental History"; Rozwadowski, "Promise of Ocean History."

22. Wertenbaker, *Floor of the Sea*, 9.

23. About technological mediation to know and perceive the oceans, see Rozwadowski and Van Keuren, *Machine in Neptune's Garden*; Camprubí, "Sonic Construction"; Höhler, "Sound Survey."

24. See Doel and Harper, *Prometheus Unleashed*, about ocean sciences in President Johnson's administration. For a contemporary analysis of the American ocean agenda, see Wenk, *Politics of the Ocean*.

25. Representative works addressing the dynamics and influence of military patronage over ocean sciences include Hamblin, *Oceanographers and the Cold War*; Oreskes, *Science on a Mission*; Weir, *An Ocean in Common*; Mukerji, *Fragile Power*; and Sapolsky, *Science and the Navy*.

26. For example, Gennesseaux and Mascle, "La naissance de la géologie marine"; Boillot, *Comment l'idée vient au géologue*.

27. See the archives of France's Ministry of Defense and its armed forces, including documents about the naval forces. Some documents indicate that partnerships went beyond academic-military relations. The Hydrographic Service shared seabed maps and data with oil companies, and embarked its experts for training the oil companies in the use of underwater technologies.

28. About the development of ocean sciences in the UK during the Cold War, see Robinson, *Ocean Science*.

29. Oreskes and LeGrand, *Plate Tectonics*; Frankel, *Continental Drift Controversy*; Le Pichon, "Fifty Years of Plate Tectonics."

30. For military patronage in the UK, see Robinson, *Ocean Science*; for military technologies in the Mediterranean Sea, see Camprubí, "No Longer an American Lake." An international perspective on developing an oceanic legal framework can be found in Ranganathan, "Ocean Floor Grab." The military-supported exploration of underwater habitats in Canada is examined in Adler, "Deep Horizons." About the relation between ocean sciences and fishing industries in Norway, see Schwach, "Sea Around Norway"; for Sweden, Lidström, Sörlin, and Svedäng present how scientific data has shown complex dynamic trends about the condition of the Swedish Seas, beyond the common declinist narrative, in their "Decline and Diversity in Swedish Seas." On the use of ocean sciences for diplomacy in France, see Martínez-Rius, "For the Benefit of All Men."

31. While France was intensively following US progress, in Japan, marine geosciences set an eye on France's progress. This creates an interesting mirror effect, in which different oceangoing nations were focusing on the capabilities (future or existing) of other countries as a motive and guideline to develop their own.

32. Quoted in a headline in *The New York Times*, November 23, 1918, p.3. For a detailed history of the oil industry, in relation to geopolitical events and changes, see Yergin, *The Prize*. For a special focus on France during the interwar period, see Nouschi, *La France et le pétrole*.

33. For the history of France's oil industry, see Nouschi, *La France et le pétrole*; Vindt, "De la CFP à Total"; Beltran, "L'industrie pétrolière en France."

34. After the fall of the Ottoman Empire, representatives from France, Great Britain, Italy, and Japan signed the San Remo Agreement (April 1920), in which the Allied Powers divided the former Ottoman Empire into administered regions. The San Remo Oil Agreement, signed by France and Great Britain, granted France a 23.75 percent share of the Turkish Petroleum Company.

35. Some works in the history of science and technology have shown the interconnection between state interests, diplomatic and geostrategic relations, and

prospecting technologies. See Cantoni, *Oil Exploration, Diplomacy, and Security* on how oil and gas prospecting in Algeria shaped diplomatic relations between the US, France, and Italy; see Camprubí, "Resource Geopolitics," for the critical diplomatic value of geophysical technologies for phosphate prospection in the Western Sahara; and Adamson, "Les Liaisons Dangereuses," for secret Franco-American collaborations with postcolonial consequences through uranium prospecting in the French Protectorate of Morocco. On the history of the geophysical company Schlumberger, and how it created a new way of doing science, see Bowker, *Science on the Run.*

36. United Nations, *Convention on the Law of the Sea,* December 10, 1982.

37. Douglas, Harper. "Etymology of 'exploitation.'" Online etymology dictionary. Accessed December 17, 2024. https://www.etymonline.com/word/exploitation.

38. Rozwadowski, *The Sea Knows No Boundaries.*

39. On military surveillance of the territory during the Cold War, see Turchetti and Roberts, *Surveillance Imperative.*

40. On the history of French bathyscaphes, see Jarry, *L'aventure des bathyscaphes.*

41. Clarke, *The Deep Range*; Clarke, *Dolphin Island*; Steinbeck, *Log from the Sea of Cortez.*

42. For the history of Greenpeace framed in wider social and political movements, see Zelko, *Make It a Green Peace!*

43. Rozwadowski, *Vast Expanses.*

44. Piketty, "Texte pour l'audiovisuelle XXème anniversaire CEPM," n.d. (ANF, 19980125/36).

45. The vessel was named as a tribute to Jean-Baptiste Charcot, the French explorer and naval officer who led the first French Antarctic Expedition (1904–7).

CHAPTER 1

1. This chapter's opening is inspired by the introductory question of Bourcart, *Géographie du fond des mers,* where the author wonders, "How can we describe the relief of the ocean floor, unknown or denied just a hundred years ago, and that we are only beginning to glimpse?"

2. Bourcart, *Géographie du fond des mers.*

3. Gougenheim, "Funérailles de Jacques Bourcart."

4. Bourcart, *Les confins albanais.*

5. Bourcart, *Les confins albanais*; for the origins of the Service Géographique de l'Armée, see Schiavon, *Itinéraires de la précision.*

6. Forces Françaises Combattantes, "Fiche individuelle de l'agent 20205" (SHD, GR 16P 81000).

7. Bourcart, *Les confins albanais,* 93, 100–119.

8. Gaston-Breton, *Total, un esprit pionnier,* 31.

9. For the history of France's oil industry, see Nouschi, *La France et le pétrole.* The company TOTAL (formerly CFP) has supported the publication of its own historical narrative (Gaston-Breton, *Total, un esprit pionnier*). Notable historical works addressing France's oil industry include Mounecif, *Chercheurs d'or noir* (about the CFP's training of engineers and experts), and Cantoni, *Oil Exploration, Diplomacy, and Security* (focusing on France's oil exploration ventures in Algeria).

10. Algeria, Tunisia, Ivory Coast, French Dahomey—current Benin, French Sudan (Mali), French Guinea, Mauritania, Niger, Senegal, French Upper Volta (Burkina Faso),

French Togoland (Togo), Chad, Oubangui-Chari (Central Africa Republic), the French Congo, Gabon, and French Cameroon.

11. In the Indian Ocean, France controlled the islands of Madagascar, Mauritius, Reunion, Mayotte, Seychelles, the Scattered Islands, and Comoros; in the South Atlantic Ocean, Sao Tomé and Príncipe; and in the Pacific Ocean, New Caledonia, French Polynesia, Wallis and Futuna.

12. Historian Radouan Mounecif has exhaustively analyzed how the CFP recruited and trained its experts from geological institutions, a key asset for the growth of the national oil company; see Mounecif, *Chercheurs d'or noir.*

13. For the work of French geologists in Morocco, exploring for the Service de la Carte Géologique, see Médioni, "L'oeuvre des géologues français au Maroc."

14. Bourcart, "Essai de classement des formations continentales du Maroc occidental"; Bourcart, "Premiers résultats"; Bourcart, "Les dépôts du second cycle miocène."

15. Bourcart, "La marge continentale."

16. Rozwadowski, *Vast Expanses.* In France, books about the ocean became popular among wide audiences in the late nineteenth and early twentieth century. Some examples would include Magny, *L'océan*; Michelet, *La mer*; or Bourée, *L'océanographie vulgarisée.*

17. The early history of echo-sounders can be found in Lawrence, *Upheaval from the Abyss*; Bates et al., *Geophysics in the Affairs of Man.*

18. On the development of sonar systems in the US Navy during wartime, see Weir, *An Ocean in Common*; Lawrence, *Upheaval from the Abyss*; Bates et al., *Geophysics in the Affairs of Man.*

19. Marti, "Rapport sur les expériences de sondage," November 12, 1919 (SHD, MV 9JJ 39).

20. Dolan, "An Early Example"; Sewell and Wiseman, "Le relief du fond de la mer."

21. Francis P. Shepard, "*Autobiography,*" unpublished manuscript (August 1980).

22. Weir, *An Ocean in Common.*

23. Curray, "*Francis P. Shepard.*"

24. Shepard, "*Autobiography.*"

25. Shepard, "American Submarine Canyons."

26. Oreskes, *Science on a Mission.*

27. Shepard, "Underlying Causes of Submarine Canyons."

28. Bourcart, "Le 'Rech' Lacaze-Duthiers."

29. Bourcart, "La marge continentale."

30. Direction Générale des Services Spéciaux, "Fiche de déclarations P. 279758," October 23, 1944 (SHD, GR 28 P58 [80]). Villat, "Mémoire de proposition" (SHD, GR 16 P 81000); Forces Françaises Combattantes, "Fiche individuelle de l'agent 20205" (SHD, GR 16P 81000); Forces Françaises Combattantes, "Questionnaire signalétique" (SHD, GR 16 P 81000).

31. Forces Françaises Combattantes, "Fiche individuelle de l'agent 20205" (SHD, GR 16P 81000).

32. Villat, "Mémoire de proposition" (SHD, GR 16P 81000); Forces Françaises Combattantes, "Questionnaire signalétique" (SHD, GR 16 P 81000).

33. Scharf, "Truman Proclamation."

34. Robelius, "*Giant Oil Fields.*"

35. Priest, "Extraction Not Creation."

36. According to Priest, the poor seismic technology used to explore oil deposits indirectly was compensated with drilling. Oil companies sought to obtain large and cheap leases to allow wide-ranging exploration programs (Priest, "Extraction Not Creation"). The first identification of a salt dome in a water-covered area, in Louisiana, was published in Rosaire and Lester, "Seismological Discovery of Vermilion Bay Salt Dome."

37. Priest, "Extraction Not Creation;" Pratt et al., *Offshore Pioneers*.

38. Scharf, "Truman Proclamation." On the freedom of foreign vessels to sail across American waters, see Suarez, *Outer Limits*.

39. Hannigan, *Geopolitics of Deep Oceans*. Oil companies were pressing the government to define a legal framework so they could acquire exploration leases. In fact, the first lease requested to exploit oil in the offshore came much earlier, in 1918, when an American citizen who claimed to have found oil 40 miles off the coast in the Gulf of Mexico requested an exploration permit from the government. The government's answer was that it did not possess jurisdiction over the seafloor (Waldock, "Legal Basis").

40. Following the Truman Proclamation and to manifest their disagreement with his decision, the state governments of Texas, Louisiana, and California halted oil exploration activities in their waters. The conflict was solved in May 1953, when the US Congress passed two agreements: the Submerged Lands Act, which granted the power to issue oil and gas leases to state governments up to 4.8 kilometers off their coastlines; and the Outer Continental Shelf Act, which situated all lands beyond 4.8 kilometers under federal jurisdiction. See National Academies of Sciences, Engineering, and Medicine, "History of the Offshore Oil and Gas Industry."

41. Churchill and Lowe, *The Law of the Sea*.

42. Buzan, *Seabed Politics*, 1; Steinberg, *Social Construction of the Ocean*, 75–86.

43. Grotius's writing can be better understood in the context of contemporary struggles between Western empires to control commercial routes: Grotius's writing aimed at vindicating the claims of his employer, the Dutch East India Company, to foster trade in the areas claimed by the Portuguese. For more, see Churchill and Lowe, *Law of the Sea*.

44. Scharf offers a great historical overview of oceanic jurisdiction before, and after, the Law of the Sea (Scharf, "Truman Proclamation").

45. Suarez, *Outer Limits*.

46. Buzan, *Seabed Politics*.

47. In Latin America, Mexico, Argentina, Panama, Chile, Peru, Costa Rica, Guatemala, Brazil, Nicaragua, Honduras and Ecuador; nine Persian Gulf sheikdoms; and Iceland, Saudi Arabia, Iran, the Philippines, Pakistan, and Yugoslavia.

48. Suarez, *Outer Limits*; Buzan, *Seabed Politics*.

49. Kreidler, *"Offshore Petroleum Industry,"* 83.

50. Priest, *Offshore Imperative*.

51. CUSS is based on the initial letters of these firms: Continental, Union, Shell, and Superior. Burleson, *Deep Challenge!*, 39.

52. Pratt et al., *Offshore Pioneers*.

53. National Academies of Sciences, Engineering, and Medicine, "History of the Offshore Oil and Gas Industry."

54. Kreidler argues that, although the offshore industry was competitive, "the complexity of marine operations required a cooperative effort" (in Kreidler, *"Offshore*

Petroleum Industry"). The Offshore Operators Committee has evolved into the oil and gas industry's principal representative regarding regulation of offshore exploration, development, and producing operations in the Gulf of Mexico.

55. Rusnak, "Afoot and Afloat Along the Edge."

56. Shepard, *Submarine Geology*, 127–28; Shepard, "*Autobiography*."

57. Shepard, "*Autobiography*."

58. Shepard, "*Autobiography*."

59. For the development and influence of military patronage on marine geosciences, see Oreskes, *Science on a Mission*; Hamblin, *Oceanographers and the Cold War*; and Weir, *An Ocean in Common*.

60. On oil exploration in Algeria after World War II, see Cantoni, *Oil Exploration, Diplomacy, and Security*. For a firsthand account, see Combaz, "Les premières découvertes de pétrole au Sahara."

61. Bourcart, "Géologie sous-marine de la baie de Villefranche," and Gougenheim, "Les canyons sous-marins."

62. The COEC did not possess its own budget, but instead relied upon the Service Hydrographique de la Marine's budget to support research, acquire materials and documents, publications, etc. See Ministre des Armées, "Arrêté—organisation et fonctionnement" (SHD, AA553 5I 283).

63. An Aviso is a relatively small, fast, and light type of military vessel. Its design prioritizes speed and agility rather than capabilities for heavy combat. This kind of ship has traditionally been used to deliver messages, perform patrol duties, or conduct scouting tasks.

64. Lalou, "Interview with Claude Lalou by Jean-François Piccard"; Gennesseaux and Mascle, "La naissance de la géologie marine."

65. G. Boillot, interview by Martinez-Rius.

66. Bourcart, "Topographie sous-marine et sédimentation."

67. Bourcart, "Hypothèses sur la genèse des gorges sous-marines."

68. Bourcart, *Géographie du fond des mers*.

69. Bourcart, "Hypothèses sur la genèse des gorges sous-marines."

70. Guilcher, "Francis P. Shepard."

71. Shepard, "*Autobiography*." On the history of the laboratory of marine geology at Villefranche's marine station, see Boillot, "1956–2012: La géologie marine."

72. Kullenberg, "Piston Core Sampler."

73. In Bourcart, "Géologie sous-marine de la baie de Villefranche." About Pettersson's expedition, see his *Westward Ho with the Albatross*.

74. Bourcart, "Colloque de géologie sous-marine."

75. Bourcart, "Géologie sous-marine de la baie de Villefranche."

76. Colloques nationaux du CNRS, *Océanographie géologique et géophysique*, 100.

77. Funded in 1949 as Lamont Geological Observatory at Columbia University, the institute was renamed in 1969 as Lamont-Doherty, and in 1993 changed from Geological Observatory to Earth Observatory. For simplicity it will generally be referred to just as Lamont.

78. Colloques Internationaux du CNRS, *La topographie et la géologie des profondeurs océaniques*; also reflected in the American publications by Fahlquist, "*Seismic Refraction Measurements*," and Gaskell and Swallow, "Seismic Refraction Experiments."

79. Colloques nationaux du CNRS, *Océanographie géologique et géophysique*, 100.

80. Bourcart, *Problèmes de géologie sous-marine*; Bourcart, *Géographie du fond des mers*.

CHAPTER 2

1. Matsen, *Jacques Cousteau*, 116.

2. See Hamblin, *Oceanographers and the Cold War*.

3. For more on the International Geophysical Year, see, e.g., Collis and Dodds, "Assault on the Unknown"; and on its organization, Korsmo, "Genesis of the International Geophysical Year."

4. Rozwadowski, *The Sea Knows No Boundaries*.

5. Although they succeeded in forging interdependent relations between distant scientific communities, part of this success can be attributed to the prevailing colonialism in many Asia-Pacific regions: most of the Asian representatives were colonial Europeans, who, therefore, shared values, ways of conducting scientific research, and mutual scientific trust.

6. Hamblin, *Oceanographers and the Cold War*.

7. Hamblin, *Oceanographers and the Cold War*.

8. Bourgoin, "Henri Lacombe, 1913–2000."

9. Historians Lino Camprubí and Sam Robinson have discussed Lacombe's role in studying the Strait of Gibraltar underwater currents in their "Gateway to Ocean Circulation." On the Mediterranean's strategic military value, see Camprubí, "No Longer an American Lake."

10. Camprubí and Robinson, "Gateway to Ocean Circulation."

11. DGRST, "Rapport du Comité d'Études Exploitation des Océans," March 12, 1960, 10 (ADF, 236QO/35).

12. Kullenberg, *Ocean Science and International Cooperation*.

13. Roberts, "Intelligence and Internationalism." The IACOMS was founded in 1956. Kullenberg, *Ocean Science and International Cooperation*; Brunn, "International Advisory Committee."

14. The report was entitled "Oceanography and Defense in the USSR, 1956–1958." Robinson details how the report was drafted after "scientific intelligence gathering" through open-source intelligence: collecting data from unclassified sources, scientific publications, and international meetings. It proved that the Soviets were the only ones possessing a deep-water magnetic oceanographic vessel, while it also revealed the size of the Soviet oceanographic fleet (Robinson, *Ocean Science*; Norman F. Ramsay, "Cooperative Oceanographic Research," April 4, 1959 (NATO, C-M[59]44).

15. Turchetti, "Sword, Shield and Buoys."

16. DGRST, "Rapport du Comité d'Études Exploitation des Océans," March 12, 1960 (ADF, 236QO/35).

17. Bourcart, *La connaissance des profondeurs océaniques*.

18. Bourcart, *Le fond des océans*.

19. DGRST, "Rapport du Comité d'Études Exploitation des Océans," March 12, 1960 (ADF, 236QO/35).

20. On the effect of declinist discourses in policymaking, see Edgerton, *Rise and Fall of the British Nation*.

21. Hamblin, *Oceanographers and the Cold War*.

22. DGRST, "Rapport du Comité d'Études Exploitation des Océans," March 12, 1960, 3 (ADF, 236QO/35); Chatry, *Il était un fois l'IFREMER*.

23. Churchill and Lowe, *Law of the Sea*. In 1948, the United Nations had constituted the International Law Commission, a body of thirty-four eminent lawyers from

different countries responsible for drafting the conventions that were discussed during the UNCLOS I. One lawyer at the International Law Commission, Georges Scelle, was French. However, experts at the International Law Commission were not meant to express/represent the official views or interests of their respective countries, but to contribute in an independent, personal capacity as legal experts.

24. Vanderpool, "Marine Science and the Law of the Sea"; Hollick, *U.S. Foreign Policy and the Law of the Sea.*

25. Enacted in the 1959 Convention on the High Seas, the 1958 Convention on the Territorial Sea and Contiguous Zone; the 1958 Convention on the Continental Shelf, and the 1958 Convention on Fishing and Conservation of the High Seas.

26. United Nations, *Convention on the High Seas 1958*, Article 2.

27. United Nations, *Convention on the Continental Shelf*, Article 1. Natural resources encompassed "mineral and other non-living resources of the seabed and subsoil together with living organisms belonging to sedentary species."

28. United Nations, *Third Meeting, 3 March 1958.*

29. Hartingh, "La position française."

30. Quéneudec, "France et le droit de la mer."

31. Sayle, *Enduring Alliance.* For an analysis framed in NATO's Science Committee, see Turchetti, *Greening the Alliance.*

32. Jacq, "Emergence of French Research Policy"; Jacq, "Aux sources de la politique de la science"; Lelong, "Le Général de Gaulle et la recherche en France."

33. Hecht, *Radiance of France*; McDougall, *"Space-Age Europe."*

34. As shown by David Edgerton (*Rise and Fall of the British Nation*), declinist discourses were common among left-wing intellectuals in the 1960s UK, who offered a biased image of the nation's past, exacerbating its economic and political failures in relation to other countries. These discourses became a powerful tool to build up particular political ideologies and agendas.

35. DGRST, "Note sur un fonds de développement de la recherche scientifique et technique," April 13, 1959 (ANF, 19920548/09). The term "undeveloped" referred to France's scientific research also appears in DGRST, "Note sur le financement de la recherche scientifique," April 16, 1959 (ANF, 19920548/9).

36. Simoncini, *Histoire de la recherche.*

37. DGRST, "Note à l'attention de messieurs les sécretaires des Comités d'Études," February 22, 1960 (ANF, 19920548/16).

38. DGRST, *Les actions concertées: Rapport d'activité 1961*, 30.

39. *Journal officiel de la République Française*, August 29, 1962: 738, 737–47.

40. Pacque, "Rapport relatif à un programme national d'océanographie," March 2, 1962 (ADF, 236QO/35).

41. DGRST, "Les actions concertées: Leur financement," June 15, 1959 (ANF, 19920548/9).

42. *Journal officiel de la République Française*, December 1959.

43. Currency conversion in 1960 US dollars, which corresponded to 42.45 million francs. It was followed by the "energy conversion" program, supported with 45.9 million francs. DGRST, "Avant-projet de répartition des crédits en 1965," August 11, 1964 (ANF, 19920548/16). By the end of 1964, the "exploitation of the oceans" had consumed 27 percent of the DGRST's special budget, which had been divided among nine high-priority research fields (here, from the most to the least funded): molecular biology, energy conversion, cancer and leukemia, demographic analysis, nutrition,

psychopharmacology, genetics, and neurophysiology. Space research also received part of the DGRST special budget.

44. DGRST, *Les actions concertées: Rapport d'activité 1961*, 30.

45. DGRST, "Rapport du Comité d'Études Exploitation des Océans," March 12, 1960 (ADF, 236QO/35).

46. DGRST, "Note sur le financement de la recherche scientifique," April 16, 1959, 2 (ANF, 19920548/9).

47. Kaldewey and Schauz, "Why Do Concepts Matter in Science Policy?"; Edgerton, "'The Linear Model' Did Not Exist."

48. Praise for fundamental research has its origins in the wake of World War II when Vannevar Bush, director of the US Office of Scientific Research and Development during the war, published the booklet *Science, The Endless Frontier* (1945), calling for the government's responsibility to support fundamental science and its role in society. On the US rhetoric of promoting fundamental sciences as a means to exert influence over Western nations, see Krige, *American Hegemony.*

49. DGRST, *Les actions concertées: Rapport d'activité 1961*, 30.

50. Pacque, "Rapport relatif à un programme national d'océanographie," March 2, 1962, 1 (ADF, 236QO/35).

51. Techniques to conduct marine geophysical studies can be classified into gravity, magnetic, heat flow, and seismic techniques, these later embracing reflection and refraction seismics. Here I will focus on seismic techniques, with particular attention to diverse devices to produce the acoustic signal.

52. The difference between reflection and refraction seismic techniques was found in the kind of waves recorded by the hydrophones: the analysis of reflected, high-frequency waves (as in reflection seismics) gave geoscientists information on the uppermost layers, but failed to offer hints about the deeper structure. The analysis of refracted waves (sent with a lower frequency from the sounding source) was especially suited to study the deepest geological structure of the oceanic crust, although it provided less precise images. After processing the signals, information materialized in seismic profiles, depictions of a seafloor's vertical cut (as if looking at the sides of a layer cake) where the layers' thickness corresponded to the velocity at which the acoustic wave traveled through each kind of sediment. Knowing this velocity allowed experts to deduce the composition and depth of each layer. For a comprehensive history of refraction seismics, see Prodehl and Mooney, *Exploring the Earth's Crust.*

53. In 1927, the Geophysical Research Exploration company conducted the first successful marine seismic survey over Louisiana's coastline, which indicated the likely existence of oil deposits; see Rosaire and Lester, "Seismological Discovery of Vermilion Bay Salt Dome," and Bates, Gaskell, and Rice, *Geophysics in the Affairs of Man.* Sternlicht, "Looking Back: Seismic Exploration," explains that marine geophysics moved to academia thanks to military technological advancements (notably in hydrophones). However, Maurice Ewing had already begun testing seismic devices at sea some years earlier.

54. Bullard, "William Maurice Ewing."

55. The story of marine geophysics at Scripps is detailed in Noble Shor, *Scripps Institution of Oceanography.* According to Raitt, after the war TNT was "readily available" through the US Navy without cost, thus he and his colleagues could utilize up to twenty kilograms of dynamite per shot at no expense (Noble Shor, *Scripps Institution of Oceanography,* 90–92).

56. Hersey, "Sedimentary Basins of the Mediterranean Sea." Dynamite explosions posed another problem: they made it impossible to send sound pulses at regular and rapid intervals as the vessel sailed forward, which was an essential feature to acquire continuous profiles (instead of acquiring data from a single location).

57. More details on the *sparker's* functioning are in Caulfield, "Predicting Sonic Pulse Shapes"; Hersey, "Continuous Reflection Profiling," 49; Knott and Hersey, "Interpretation of High-Resolution Echo-Sounding Techniques."

58. Edgerton, "The 'Boomer' Sonar Source"; Knott and Hersey, "Interpretation of High-Resolution Echo-Sounding Techniques."

59. Ewing and Press, "Geophysical Contrasts." On average, oceanic crust is four times thinner than continental crust.

60. They first published their evidence in Heezen, Tharp, and Ewing, "Floors of the Oceans." This story has been recounted from different perspectives, as it constitutes a central part of the history of the discovery of plate tectonics. For an approach focused on how Tharp and Heezen dodged military secrecy, see Oreskes, "Iron Curtain of Classification." For a focus on military patronage, see Doel et al., "Extending Modern Cartography." For details on the controversy and the production of scientific knowledge, see Frankel, *Continental Drift Controversy*. For a biographical approach to Tharp, see Felt, *Soundings*.

61. Le Pichon, "Fifty Years of Plate Tectonics." Oreskes explains that Ewing was primarily interested in expanding ridges for their potential connection to the formation of underwater canyons, and not as interested relative to the still unproved continental drift (Oreskes, "Iron Curtain of Classification").

62. Hersey, "Sedimentary Basins of the Mediterranean Sea." In this paper, Hersey also acknowledges NATO's support in surveying the region, although there is no direct evidence to support the hypothesis that Hersey was conducting military research while undertaking his marine geophysical surveys. American oceanographers could rely on a logistical base in the Mediterranean because NATO had just established a SACLANT Antisubmarine Warfare Research Center at the Italian village of La Spezia, less than two hundred kilometers from Monaco. For more about US scientific-military activities in the western Mediterranean, see Ross, "Twenty Years of Research at the SACLANT ASW Research Center."

63. Jacques Cousteau expressed it during the 1961 national conference on marine geology and geophysics (Muraour, Leenhardt, and Merle, "Éléments pour un programme de recherches séismiques").

64. Schiefelbein and Cousteau, *The Human, the Orchid, and the Octopus*, refers on numerous occasions to Cousteau's relation to marine geosciences, including references to the development of the Mediterranean's seafloor geology and his contributions with Edgerton. Axel Madsen, *Cousteau: An Unauthorized Biography* makes references to his relationship to scientists and the joint missions organized onboard *Calypso*; Madsen also makes a brief reference to marine geophysical surveys onboard the *ship*, labeling them "bombing the ocean" (124). In *Jacques Cousteau: The Sea King*, Bradford Matsen details Cousteau's relationship to Edgerton and the development of the *boomer* for marine geophysics.

65. The region between the Gulf of Geneva in the north, Monaco in the west, Italy in the east, and Corsica in the south. Alinat, Giermann, Leenhardt, "Reconnaissance sismique des accidents de terrain."

66. In the original, "accidents de terrain," which translates as landforms or geomorphic features. Cousteau's label simply aimed at pointing to the unknown processes that had originated small seamounts in an area where they were not expected.

67. Bourcart, "Projets de développement des études de géologie et géophysique marines," May 9, 1961 (ADF, 236QO/35).

68. Muraour, Merle, and Ducrot, "Observations sur le plateau continental"; DGRST, *Les actions concertées: Rapport d'activité 1961*, 48–49; COMEXO, "Commission Méditerranée," November 7, 1962 (ADF, 236QO/35).

69. COMEXO, "Marche de convention protocole: Institut Océanographique Monaco," n.d. (ANF, 201103801/01). Muraour, Leenhardt, and Merle, "Eléments pour un programme de recherches séismiques."

70. Muraour, Merle, and Ducrot, "Observations sur le plateau continental"; DGRST, *Les actions concertées: Rapport d'activité 1961*, 34.

71. Most of the participants were oceanographers (mainly British and American, followed by Dutch, German, and French), but it was also attended by experts from oil companies (American Overseas Petroleum Ltd., British Petroleum Co., Shell Petroleum Co., Royal Dutch Shell, and the French SNPA), and representatives of the British Royal Navy (see Whittard and Bradshaw, *Submarine Geology and Geophysics*).

72. Hersey, "Sedimentary Basins of the Mediterranean Sea."

73. Menard et al., "Rhône Deep-Sea Fan."

74. Ewing and Antoine, "New Seismic Data."

CHAPTER 3

1. DGRST, "Projet d'exposé des motifs instituant un Centre National d'Études Océanographiques," February 10, 1964 (ANF, 19920548/19).

2. Piketty, "Texte pour l'audiovisuelle XXème anniversaire CEPM," n.d. (ANF, 19980125/36).

3. Cherruau and Ferrari, "André Giraud."

4. The Corps des Mines, a prestigious French administrative and engineering body, played a key role in managing and advising on matters related to industry, energy production, and mining.

5. Starr, "André Y. Giraud 1925–1997."

6. Giraud and Boy de la Tour, *Géopolitique du pétrole et du gaz*.

7. At the same time, global oil consumption skyrocketed: from 469 million tons in 1949, to 763 million tons in 1955, and 1,051 million tons by 1960 (Nouschi, *Pétrole et relations internationales*, 35).

8. Salut, "Politique nationale du pétrole."

9. Gaston-Breton, *Total, un esprit pionnier*.

10. Nyaberg, "A Few Strategic Considerations."

11. Salut, "Politique nationale du pétrole."

12. In Piketty, "Allocution de Gérard Piketty."

13. Data from the US Energy Information Administration, consulted in https://www.eia.gov/totalenergy/data/annual/showtext.php?t=ptb0502. The history of offshore oil exploration in the US appears in Burleson, *Deep Challenge!* Other authors have tackled the development of particular companies in the offshore, including Pratt et al., *Offshore Pioneers*, and Priest, *Offshore Imperative*.

14. Burleson, *Deep Challenge!* Oil exploration in the North Sea has received notable scholarly attention, e.g., Kemp, *Official History of North Sea Oil and Gas*, and Harvie, *Fool's Gold.*

15. Chapelle, "Le pétrole et la mer."

16. Giraud, "Note sur la création d'un établissement public pour le développement de l'océanographie et la mise en valeur de la mer," February 28, 1966 (ANF, 20060160/1).

17. The term *neocolonialism* applied to the oceans appears in contemporary discussions, especially related to the new ocean legislation. See Young, "Legal Regime," and Robinson, "Scientific Imaginaries and Science Diplomacy."

18. Piketty, "Texte pour l'audiovisuelle, XXème anniversaire CEPM," n.d. (ANF, 19980125/36).

19. CEPM was first named *Comité d'Études Marines*. The Régie Autonome des Pétroles (RAP) merged in 1965 with the Bureau de Recherche de Pétrole (BRP) to form the Entreprise de Recherches et d'Activités Pétrolières (ERAP). Leblond, "Décision du directeur des carburants," May 7, 1963 (ANF, 20150386/156).

20. Leblond was CEPM's director during the initial months, before Giraud replaced him.

21. Combaz, "Le comité d'études pétrolières marines." CEPM's structure and goals were not unprecedented in France. As Roberto Cantoni has detailed (*Oil Exploration, Diplomacy, and Security*, 85–88), the Bureau de Recherche de Pétrole was created in 1945 to coordinate and fund oil exploration abroad.

22. Chapelle "Le pétrole et la mer."

23. "Note pour Mr. Amiral Chef d'Etat-Major de la Marine. Autorisation d'embarquement sur un bâtiment hydrographique," July 2, 1963 (SHD, MV 9 JJ 567).

24. Georges Cabanier, "Compte-rendu. Collaboration entre la marine et les organismes pétroliers: Réunion du 28 juin 1966," September 28, 1966 (SHD, MV 9 JJ 567).

25. Maurice Leblond, letter to the IFP general director, June 17, 1963 (ANF, 20150386/108); and André Giraud, letters to the IFP general director, February 8, 1964, and February 20, 1965 (ANF, 20150386/108).

26. Equivalent to 42.45 million francs.

27. McQuillin, *Introduction to Seismic Interpretation*, 40.

28. On the environmental controversy between fishermen and the oil industry, see Locher, "Lutter contre l'empire du pétrole."

29. ERAP, "*Programme d'études marines 1966*" (AHT, 07AH119/76); Yvonne Rebeyrol, "Le Flexotir ou comment chercher du pétrole en mer sans tuer les poissions." *Le Monde*, October 17, 1966. As noted above, ERAP was born in 1965 from the merging of RAP and the Bureau de Recherche de Pétrole (BRP).

30. Hersey, "Continuous Reflection Profiling."

31. Equivalent to 1.5 million francs. In André Giraud, letter to the IFP general director, February 8, 1964 (ANF, 20150386/108).

32. IFP, *Rapport annuel 1963.*

33. Cholet and Fail, "Dispositif émetteur d'ondes sonores;" Cholet and Fail, "Patent Specification 1.097.227."

34. Rebeyrol, "Le Flexotir."

35. Cassand et al., "Sismique Réflexion en Eau Profonde (Flexotir)."

36. ERAP, "*Programme d'études marines 1966*" (AHT, 07AH119/76); ERAP, "Programme d'études marines: Rapport annuel 1968" (AHT, 07AH119/76) ; IFP, *Rapport annuel 1966, 1967, 1968, 1969, 1970.*

37. *Journal officiel de la République Française*, August 29, 1962, 742. American oceanographer John B. Hersey would point to the existence of salt domes in the deep Mediterranean soon after, as shown in chap. 2.

38. McQuillin, *Introduction to Seismic Interpretation.*

39. Delacour, Parola, and Grolet. "Dispositif pour le carottage sous-marine."

40. Delacour and Castela, "Appareillage de forage sous-marin."

41. In 1964, the IFP's Division of Geophysics worked at sea 121 out of 200 days (IFP, *Rapport annuel 1964*). This exceeded the days at sea spent by academic geologists and geophysicists, who could not undertake a complex mission until 1966, when the *Jean Charcot* was set in operation (IFP, *Rapport annuel 1965*).

42. G. Boillot, interview by Martinez-Rius. Also in Boillot, *Comment l'idée vient au géologue.*

43. ERAP, *Rapport annuel 1968* (AHT, 07AH119/76). Another case, involving the University of Montpellier and the sonobuoys there developed, is detailed at ERAP, "Programme d'études marines: Rapport annuel 1966" (AHT, 07AH119/76).

44. DGRST, "Projet d'exposé des motifs instituant un Centre National d'Études Océanographiques," February 10, 1964 (ANF, 19920548/19).

45. For President Johnson's interest in promoting oceanography, see Doel and Harper, "Prometheus Unleashed," and Adler, "Cold War Science on the Seafloor." A detailed history of oceanography under Johnson's presidency can be found in Hamblin, *Oceanographers and the Cold War.*

46. Hamblin, *Oceanographers and the Cold War, xvii.*

47. For a firsthand account of the creation of the commission, see Merrell et al., "Stratton Commission," and Wenk, *Politics of the Ocean.*

48. "Un milliard de francs pour l'océanographie américain" (ANF, 20060160/1); André Giraud, letter to Raymond Toussaint, February 28, 1966 (ANF, 20060160/1).

49. Comité Consultatif de la Recherche Scientifique et Technique, *Rapport au Comité Interministériel de la Recherche Scientifique et Technique*, March, 29, 1966 (ANF, 20060160/1); DGRST, "Note sur l'organisation de l'océanographie," November 29, 1965 (ANF, 19920548/19); and André Marechal, letter to Maurice Fontaine, January 20, 1966 (ANF, 20060160/1).

50. In a talk about the past and present of the French National Research Council (CNRS), historian Dominique Pestre highlighted that the use of "valorization" to talk about the promoted science might be a euphemism to emphasize the economic returns that scientific research was expected to provide. However, he does not question the term itself, but only its use and contextual meaning (as he also does with "fundamental" and "multidisciplinary"). This probably exemplifies how integrated the term is within the French language. Pestre, "L'évolution des champs de savoir."

51. "R/V" stands for Research Vessel, a ship equipped and used for scientific research at sea.

52. Giraud, "Note sur la création d'un établissement public"; DGRST, "Projet d'exposé des motifs instituant un Centre National d'Études Océanographiques," February 10, 1964 (ANF, 19920548/19).

53. André Marechal, letter to Maurice Fontaine, André Giraud, and Jean-Marie Pérès, "À propos du COMEXO II," January 20, 1966 (ANF, 20060160/1).

54. DGRST, "Note sur le projet de création d'un Centre National d'Études Océanographiques," October 5, 1964 (ANF, 20060160/1).
55. COMEXO, "Procès-verbal de la 2ème réunion du COMEXO," February 21, 1966 (ANF, 20060160/1).
56. DGRST, "Note sur le projet de création."
57. Comité Consultatif de la Recherche Scientifique et Technique, *Rapport au Comité Interministériel de la Recherche Scientifique et Technique*, January 29, 1966, 57–59 (ANF, 20060160/1).
58. Cousteau and Jacquier, *Français, on a volé ta mer.*
59. DGRST, "Projet d'exposé des motifs instituant un Centre National d'Études Océanographiques," February 20, 1964 (ANF, 19920548/19).
60. In DGRST, "Note sur l'organisation de l'océanographie," November 29, 1965 (ANF, 19920548/19). More detail on CNES, and its similarities with the forthcoming CNEXO, in chap. 5.
61. DGRST, "Note sur l'organisation de l'océanographie," November 29, 1965 (ANF, 19920548/19).
62. André Giraud, letter to the president of the *Comité Consultatif de la Recherche Scientifique et Technique*, April 12, 1966 (ANF, 20060160/1); André Giraud, letter to Yves la Prairie, September 26, 1966 (ANF, 20060160/1).
63. DGRST, "Création d'un organisme de coordination de la recherche océanographique," June 1966 (ANF, 20060160/1).
64. Ministre délègue de la Recherche Scientifique et Technique et des Questions Spatiales, to Mr. Ministry [*sic*] of Industry (André Giraud), June 1966 (ANF, 20060160/1, Documents André Giraud—dossier océanographie).
65. As la Prairie mentions in his memoirs, *Ce siècle avait de Gaulle.*
66. La Prairie, *Ce siècle avait de Gaulle.*
67. Hecht, *Radiance of France.*
68. Historian Jacob Hamblin suggests one more reason to justify why an expert outside the discipline might be selected to drive oceanography. In the US, the harsh rivalry between oceanographic institutions forced the National Academy of Sciences to choose a geochemist, Harrison Brown, to implement a national oceanographic program, rather than one of the directors of the main institutions (Hamblin, *Oceanographers and the Cold War*, 141–43).
69. La Prairie, *Ce siècle avait de Gaulle*, 326.
70. DGRST, "Note sur le projet de création."
71. This event is also described by la Prairie in his memoirs. Peyrefitte crossed off the name of Giraud, while Jacques Cousteau, also a candidate for the position, rejected the proposal by claiming that he'd rather be a "free wolf" than a "pet dog serving the state" (la Prairie, *Ce siècle avait de Gaulle*, 349).
72. On the role of scientists as political advisors, see Mukerji, *Fragile Power.*

CHAPTER 4

1. Yvonne Rebeyrol, "Un essai de coopération recherche-industrie: La prochaine campagne du *Jean-Charcot.*" *Le Monde*, April 17, 1968.
2. Yves la Prairie, letter to André Giraud, May 22, 1968 (ANF, 20160129/327). Giraud also aimed at transferring some of the coordination from CEPM to CNEXO, especially the projects related to "penetration of men underwater," "seawater corrosion," and

"action of natural elements on materials"; Yves la Prairie, letter to André Giraud, March 21, 1968 (ANF, 20160129/327).

3. United Nations, *Convention on the Continental Shelf*. In France, conditions were drafted in CNEXO, "Projet de loi et projet de décret relatifs à l'exploration et l'exploitation du plateau continental," June 17, 1968 (20060160/1).

4. United Nations, *Convention on the Continental Shelf*, Article 5.8.

5. Rebeyrol, "Un essai de coopération recherche-industrie."

6. Yves la Prairie, "Fiche—campagne du navire océanographique *Jean Charcot* dans le Golfe de Guinée en 1968," May 22, 1968 (ANF, 20160129/327).

7. Rebeyrol, "Un essai de coopération recherche-industrie."

8. Yves la Prairie, letter to André Giraud, June 22, 1968 (ANF, 20160129/327).

9. Fail et al., "Prolongation des zones de fractures."

10. For detail on this and later campaigns, see Schlich, *BENIN 1971 cruise, RV Jean Charcot*. It is likely that CNEXO focused on that region after the American campaign onboard the *Discoverer* collected large amounts of data there, supported by the US Geodetic Survey (Vanney and Pinot, "La campagne du *Discoverer*").

11. The team did not mention the phrase "plate tectonics," since the theory had just been proposed (Fail et al., "Prolongation des zones de fractures").

12. Wilson, "A New Class of Faults."

13. He relied on data acquired by the World-Wide Standardized Seismograph Network of the US Coast and Geodetic Survey, installed for military purposes. Sykes, "Mechanism of Earthquakes."

14. Le Pichon, "Sea-Floor Spreading and Continental Drift."

15. Oreskes and LeGrand, *Plate Tectonics*; Le Pichon, "Fifty Years of Plate Tectonics."

16. Fail et al., "Prolongation des zones de fractures."

17. Yves la Prairie, letter to André Giraud, May 22, 1968 (ANF, 20160129/327).

18. CNEXO, Contrat 08/163, "Procédé FLEXOTIR," June 28, 1969 (ANF, 20110381/10). If CNEXO did not comply with these conditions, the contract specified that it must pay to the IFP 25 French francs (the equivalent of 0.05 USD of 1969) per kilometer of seismic profile acquired.

19. Oil interest expressed in "Rapport pour l'Assemblée Nationale—Questionnaire pour la préparation de la loi de finances pour 1972," n.d. (ANF, 20160259/21).

20. Le Pichon, "Sea-Floor Spreading and Continental Drift."

21. La Prairie specified his interest in le Pichon's American experience in, e.g., his letter to M. Parker (technical advisor at the ministry of scientific research), June 26, 1968 (ANF, 20160129/327); and in his letter to the associate director of the Lamont Earth Observatory, J. L. Worzel, March 28, 1968 (ANF, 20160129/327).

22. Yves la Prairie, letters to Xavier le Pichon, January 11, 1968, and February 7, 1968 (both in ANF, 20160129/327).

23. Le Pichon, *Kaiko*.

24. G. Boillot, interview by Martinez-Rius. Noratlante I was designed as a multidisciplinary campaign in marine geosciences and marine biology, led by the two CNEXO advisors, le Pichon and marine biologist Lucien Laubier. With this mission, both aimed at organizing something unprecedented for France (see le Pichon, *Kaiko*).

25. Groupe scientifique du COB, *Résultats des campagnes*.

26. Yvonne Rebeyrol, "L'opération Noratlante." *Le Monde*, August 5, 1969.

27. Pautot, "La dorsale médio-atlantique"; Olivet et al., "La faille transformante Gibbs."

28. Pautot et al., "Continuous Deep Sea Salt Layer"; le Pichon et al., "Geophysical Study."

29. This strategy was stated in numerous documents, including le Pichon, "Mission Nestlante I."

30. La Prairie promoted a policy of data-sharing, following the example of other advanced oceanographic countries (including the US) and their international projects.

31. Le Pichon, "Mission Nestlante I."

32. Gaston-Breton, *Total, un esprit pionnier*; Mounecif, *Chercheurs d'or noir*. Elf-ERAP was the public name of a French state-owned oil group created in 1965. Legally established as ERAP (Company for Petroleum Research and Activities), it consolidated national interests by merging RAP, SNPA, and BRP. The name Elf—from Essences et Lubrifiants de France ("Petrol and lubricants of France")—was adopted as its commercial brand.

33. CNEXO, "Protocole de subvention pour la réalisation de la campagne océanographique Nestlante II," June 18, 1970 (ANF, 20110381/12).

34. CNEXO, "Commission de Finances: Développement industriel et scientifique de la recherche océanographique" (ANF, 20160129/323).

35. CNEXO, "Protocole de subvention pour la réalisation de la campagne océanographique Nestlante II," June 18, 1970 (ANF, 20110381/12).

36. Beuzart, *Nestlante II*.

37. CNEXO, *Rapport annuel 1970*; Malod and Mascle, "Structures géologiques"; Briseid and Mascle, "Structure de la marge continentale."

38. I don't have evidence that oil companies hindered the publication of any piece of data. CNEXO, "Protocole de subvention pour la réalisation de la campagne océanographique Nestlante II," June 18, 1970 (ANF, 20110381/12).

39. CNEXO, "Note préparatoire à l'établissement d'un programme CNEXO concernant l'exploitation minière des océans" (ANF, 19980125/01).

40. For a detailed analysis of oceanographers as a specialized labor workforce, see Mukerji, *Fragile Power*.

41. La Prairie, "Instructions du directeur genéral du CNEXO pour M. le Faucheaux," July 24, 1967 (ANF, 20160129/327); La Prairie, letter to the French prime minister, November 21, 1967 (ANF, 20160129/327).

42. Yves la Prairie, letter to the minister of foreign affairs, March 5, 1968 (20160129/326).

43. In a private note to his technical advisor, Olivier le Faucheaux, la Prairie declared his ambitions to receive an income from the taxes obtained through mining leases on the continental shelf. However, he didn't mention it in his subsequent letter to the prime minister, nor did it materialize in the final law (in la Prairie, "Instructions du directeur genéral du CNEXO pour M. le Faucheaux," July 24, 1967 (ANF, 20160129/327). La Prairie, letter to prime minister, November 21, 1967 (ANF, 20160129/327).

44. "Loi n° 68–1181 du 30 décembre 1968."

45. Minister of industrial and scientific development, letter to Yves la Prairie, May 11, 1971 (ANF, 20160259/321); Yves la Prairie, letter to the minister of industrial and scientific development, June 18, 1971 (ANF, 20160259/321); Yves la Prairie, letter to minister of industrial and scientific development, March 21, 1972 (ANF, 20160259/320);

La Prairie to minister of industrial and scientific development, February 4, 1972 (ANF, 20160259/320).

46. CNEXO, "Note préparatoire à l'établissement d'un programme CNEXO."

47. CNEXO, "Rapport du groupe exploitation des matières minérales et fossiles," n.d. (ANF, 19980125/01); Conseil d'Administration CNEXO, "Procès-verbal de la séance du jeudi 4 avril 1968," April 4, 1968 (ANF, 19980125/01).

48. CNEXO, "Note préparatoire à l'établissement d'un programme CNEXO"; Boillot, "Des marges continentales atlantiques."

49. This was 587,000 francs. In CNEXO, "Reconnaissance géologique du plateau continental atlantique dans le Golfe de Gascogne," contract 69/141, December 12, 1969 (ANF, 20110381/10); Maurice Vigneaux, "Rapport préliminaire concernant les activités du Laboratoire de Géophysique Appliquée," convention no. 69/141, n.d. (ANF, 20110381/10).

50. CNEXO, "Études sismiques de la marge continentale atlantique," contract 68/33, 1968 (ANF, 20110381/07); CNEXO, "Étude sismique de la marge continentale atlantique," contract 6700534, 1967 (ANF, 20110381/05).

51. CNEXO, "Reconnaissance géologique en manche orientale," contract 70/181 with Geotechnip, April 8, 1970 (ANF, 20110381/12); Gilbert Boillot, "Étude stratigraphique de carottes de roche prélevées en manche orientale," November 5, 1970 (ANF, 20110381/09); CNEXO, "Reconnaissance par carottages de surface du substratum de la manche centrale et orientale," contract 69/79 (ANF, 20110381/09).

52. CNEXO, "Synthèse des principales activités océanologiques du 1 février au 15 avril 1969," n.d. (ANF, 19980125/01). Also in CNEXO, "Synthèse des principales activités océanologiques du 15 avril au 15 juin 1969," n/d. (ANF, 19980125/01). They also offered the *Petite Marie Françoise*, their research vessel adapted to conduct marine geophysical surveys.

53. Maurice Vigneaux, "Rapport préliminaire concernant les activités du Laboratoire de Géophysique Appliquée à l'Océanographie," 1969 (ANF, 20110381/10).

54. Where the symposium was going to take place created friction between CNEXO and the IFP: according to la Prairie, organizing such a pivotal meeting at Rueil-Malmaison favored the IFP, which should thus return the favor on a future occasion.

55. Chairs of the sessions: Burollet (CFP), Drake (Dartmouth College, US); Goguel (BGM), Levy (ELF-ERAP), Matthews (Cambridge), Radier (SNPA).

56. Debyser et al., *Histoire structurale du Golfe de Gascogne*.

57. G. Boillot, interview by Martinez-Rius.

58. Le Pichon et al., "Une hypothèse d'évolution tectonique."

59. Boillot, "Des marges continentales atlantiques."

60. G. Boillot, interview by Martinez-Rius.

61. Yves la Prairie, letter to M. Fenning, January 21, 1969 (ANF, 20160129/326).

62. Yves la Prairie, letter to M. Parker, January 29, 1969 (ANF, 20160129/326).

63. Asserted by the Scientific and Technical Committee and, later, mentioned at the Administrative Council (Comité Scientifique et Technique, "Procès-Verbal de la 5ème reunion," December 18, 1968 (ANF, 20110381/05).

64. Comité Scientifique et Technique, "Procès-Verbal de la 5ème réunion," December 18, 1968 (ANF, 20110381/05).

65. CNEXO, "Programme national d'océanographie—Les choix du CNEXO," April 4, 1968 (ANF, 19980125/01).

66. CNEXO, "Synthèse des principales activités océanologiques du 15 mars 1971 au 1er juin 1971" (ANF, 19980125/02).

67. CNEXO, *Rapport annuel 1972.*

CHAPTER 5

1. Speech at the First UN General Assembly by Ambassador Arvid Pardo, Malta's delegate (UN General Assembly, First Committee Debate, November 1, 1967).

2. UN General Assembly, First Committee Debate, November 1, 1967.

3. UN General Assembly, First Committee Debate, November 1, 1967; also analyzed in Ranganathan, "Decolonization and International Law." Political scientist Surabhi Ranganathan has argued for a "grab of the seabed" during the Third UNCLOS in Ranganathan, "Ocean Floor Grab."

4. UN General Assembly, First Committee Debate, November 1, 1967. In Gorove, "Concept of 'Common Heritage,'" the author mentions the Antarctic Treaty (1959), the Non-Proliferation Treaty (1968), and the international regulation of outer space at the UN General Assembly (1959).

5. These dynamics are described, e.g., in Robinson, "Scientific Imaginaries and Science Diplomacy."

6. Hamblin, *Oceanographers in the Cold War.*

7. See www.ourworldindata.org. This exponential growth and its consequences as part of the Anthropocene are detailed in Bonneuil and Fressoz, *Shock of the Anthropocene.*

8. Hardin, "Tragedy of the Commons." First introduced in a speech for the Pacific Division of the American Association for the Advancement of Science (an audience constituted of scientists), the concept reached wider dissemination thanks to its later publication in the journal *Science.* Historian Fabien Locher has traced the ideological, intellectual, and political currents that informed the concept in Locher, "Les pâturages de la guerre froide."

9. Ranganathan has compared the opposite backgrounds and ideologies of Pardo and Hardin as well as how those informed their claims on the "commons," in Ranganathan, "Global Commons."

10. UN General Assembly, "Note verbal from the Permanent Mission of Malta," August 18, 1967.

11. UN General Assembly, First Committee Debate, November 1, 1967.

12. Argued by Hannigan, *Geopolitics of Deep Oceans.* Historian Maria Gavouneli has highlighted how Pardo's ideas fit in the ideological economic framework of his times. After World War II, former colonial systems in African and Asian territories collapsed, being replaced by a neoliberal system that increased socioeconomic differences between countries with different technological and economic capabilities (Gavouneli, "From Uniformity to Fragmentation").

13. Hannigan, *Geopolitics of Deep Oceans.*

14. Barkenbus, "Politics of Ocean Resource Exploration." The North-South battle has been pointed out as one of the characteristic features of UNCLOS III; it contrasted sharply with the classical East-West clash that pervaded UNCLOS I and II (for more, see, e.g., Miles, *Global Ocean Politics*).

15. Maurice Schumann, letter to Yves la Prairie, July 31, 1968 (ANF, 20160129/327); Yves la Prairie, letter to the minister of foreign affairs, June 13, 1968 (ANF 20160 129/327).

16. Yves la Prairie, letter to Alain Sciard, June 29, 1967 (ANF, 20160129/328).

17. Lacharrière, "Georgette Mariani."

18. Hints in the archival record point out that CNEXO played a crucial role in informing the national position at UN debates around the ocean legislation. The topic deserves attention on its own, as it can deepen understanding of the relations between science and international legislation for the oceans. See Georgette Mariani, "Schéma de discussion à propos de l'étude du Sécretariat Général des Nations Unies sur un mécanisme International," June 13, 1969 (ANF, 20160129/325); Yves la Prairie, letter to the minister of foreign affairs, November 24, 1969 (ANF, 20160129/325).

19. Political scientist Pierre-Bruno Ruffini has coined the term "science-diplomacy nexus" to conceptualize the coconstruction and relations between international policies and science (Ruffini, "Intergovernmental Panel").

20. Formally named the Committee on the Peaceful Uses of the Seabed and the Ocean Floor Beyond the Limits of National Jurisdiction, organized between 1967 and 1973.

21. Specified in the UN's Declaration of Principles Governing the Sea-bed and the Ocean Floor, and the Subsoil Thereof, Beyond the Limits of National Jurisdiction. (25th Session, 1970).

22. UN General Assembly, Resolution 2749 (XXV) of December 17, 1970.

23. Winthrop Haight, "United Nations Affairs."

24. Robinson, "Scientific Imaginaries and Science Diplomacy."

25. Jacques Perrot, Conference "Aspects économiques de l'exploitation des océans," April 1970 (ANF, 20160259/323). On how competition was articulated with cooperation, see Robinson, "Early Twentieth-Century Ocean Science Diplomacy"; Martínez-Rius, "For the Benefit of All Men."

26. Explicitly stated in "Programme national d'océanographie—chapitre 3: Les choix du CNEXO," n.d. (ANF, 19980125/1). On avoiding military or commercial implications in international cooperation, see Yves la Prairie, confidential letter to the minister of scientific research, November 29, 1967 (ANF, 20160129/328).

27. Yves la Prairie, letter to Alain Sciard, June 29, 1967 (ANF, 20160129/328). Just to mention reports from French scientific attachés: Ambassade de France à Bonn, "Rapport océanologie," June 1970 (ANF, 20080658/23); Mission scientifique— Ambassade de France en Washington, "Les nouvelles structures de l'océanographie américaine," September 14, 1971 (20080658/22) and "L'enseignement des disciplines océanographiques aux États-Unis," July 1, 1970 (20080658/22); "L'océanographie en Italie," n.d. (ANF, 20080658/23).

28. DGRST, "Note sur le projet de création d'un Centre National d'Etudes Océanographiques," October 5, 1964 (ANF, 19920548/19).

29. Jacques Perrot, letter to the minister of scientific research, February 13, 1968 (ANF, 20160129/327); Yves la Prairie, confidential note to the minister of scientific research, January 31, 1969 (ANF, 20160129/326). Also discussed in Conseil Administration CNEXO, "Procès-verbal de la 2ème réunion," October 5, 1967 (ANF, 19980125/1).

30. DGRST, "Note sur le projet de création." The budget invested by these countries appears detailed in Yves la Prairie, "Fiche sur l'effort français en océanologie," January 12, 1970 (ANF, 20160129/325).

31. On ambiguous Franco-American relations, between cooperation and competition in marine sciences, see Martínez-Rius, "For the Benefit of All Men."

32. Yves la Prairie, letter to Edward Wenk, December 20, 1967 (ANF, 20160129/328). On American ocean policy, see Wenk, *Politics of the Ocean*. On how Johnson's administration sought to use environmental sciences as a tool for foreign policy (notably oceanography and meteorology), see Doel and Harper, "Prometheus Unleashed."

33. Wenk wrote a number of books recounting his experience and learning during the time he occupied government-related positions: see his *Making Waves* and *Politics of the Ocean*.

34. Yves la Prairie, letter to Edward Wenk, December 20, 1967 (ANF, 20160129/328); Yves la Prairie, letter to M. Jean-Louis Chaussende (deputy chef to Maurice Schumann, ministry of scientific research), October 17, 1967 (ANF, 20160129/328).

35. Yves la Prairie, letter to Edward Wenk, December 20, 1967 (ANF, 20160129/328).

36. Yves la Prairie, letter to the minister of scientific research, March 20, 1968 (ANF, 20160129/327).

37. Yves la Prairie, letter to J. L. Worzel, March 8, 1968 (ANF, 20160129/327).

38. The COB was built between 1968 and 1971 on a two-million-dollar budget (CNEXO, "Fiche sur le futur Centre d'Océanologie de Brest," n.d. [ANF, 20160129/328]).

39. Another reason to build the center in Brest was the proximity of the Navy's Hydrographic Service and the numerous fishing industries that the city hosted. According to CNEXO officers, the center would also stimulate collaborations with those domains.

40. Hardy, "Letter from Paris: Exploring the Shelf."

41. These included "SIO Growth Plan: Fall 1968 Space Assignment Summary," "Permanent Buildings of the Scripps Institution of Oceanography," "Breakdown of Campus Space—Scripps Institution of Oceanography; SIO on Campus Space—Fall, 1966; revised 1968," and the hand-drawn map "Lamont Observatory, echelle 1/2200." Folder "Centres de Recherche 1969" (ANF, 20080658/22).

42. "Partie adopte pour la construction du Centre Oceanologique de Bretagne," n.d. (ANF, 20080658/22).

43. "Tentative de comparaison entre le COB et certains centres etrangers," April 27, 1970 (ANF, 20080658/22).

44. CNEXO's control of these technologies generated frictions with research teams at universities; see Laubier, "L'émergence de l'océanographie au CNRS."

45. Notably characteristic of the Lamont under the leadership of Maurice Ewing.

46. Yves la Prairie, letter to the minister of scientific research, March 20, 1968 (ANF, 20160129/327); Mission scientifique de l'ambassade de France en Washington, "Evolution de la coopération scientifique et technique Franco-Américain," n.d. (ANF, 20080658/22).

47. Edward Wenk Jr., "Terms of Reference for Marine Science Cooperation Between the National Center for the Exploration of the Oceans of France and the National Council on Marine Resources and Engineering Development of the United States of America," in Edward J. Wenk, letter to Yves la Prairie, January 20, 1970 (ANF, 20080658/22). The discourse of "the benefit of all men" was rather common

in that period. Given the public refusal of the Vietnam War and East-West tensions, the American administration was reluctant to frame any other venture as a heated competition against other nations (Hamblin, *Oceanographers and the Cold War*, 244).

48. Robinson, "Scientific Imaginaries and Ocean Diplomacy," analyzes the same will but from the American perspective, where policymakers and oceanographers interested in marine exploitation were eager to be the ones who would "conquer the oceans" first.

49. Alain Sciard, "Réflexions et suggestions sur la politique de coopération internationale du CNEXO," September 16, 1971 (20160129/321).

50. Some key papers on magnetic stripes in the seafloor include Vine and Matthews, "Magnetic Anomalies over Oceanic Ridges," and Heirtzler et al., "Magnetic Anomalies over the Reykjanes Ridge." On seafloor spreading and continental drift, see Dietz, "Continent and Ocean Basin Evolution"; Heezen, "Rift in the Ocean Floor"; and Wilson, "A New Class of Faults." On plate tectonics as a unifying theory, see Morgan, "Rises, Trenches, Great Faults," and le Pichon, "Sea-Floor Spreading."

51. For a firsthand account of the FAMOUS expedition, see Riffaud and le Pichon, *Expédition FAMOUS*.

52. Naomi Oreskes has explored military patronage relations through the military and scientific operation of *Alvin* in her "Context of Motivation."

53. Heirtzler and Grassle, "Deep-Sea Research."

54. Le Pichon, *Kaiko*.

55. Naomi Oreskes has presented the project FAMOUS from the American perspective in her "Context of Motivation." The project's organization from the US side has been described in Ballard, "History of Woods Hole's Deep Submergence Program," and, from the French perspective, in Jarry, *L'aventure des bathyscaphes*.

56. Alain Sciard, "Réflexions et suggestions sur la politique de coopération internationale du CNEXO," September 16, 1971 (ANF, 20160129/321).

57. Constituted by Jean Francheteau (COB geophysicist), Gilbert Bellaiche (sedimentologist from Villefranche-sur-Mer), Jean-Louis Cheminée (volcanologist at the CNRS), and Pierre Choukroune (geologist at the University of Rennes). Yves La Prairie, "Note pour Monsieur le Contrôleur d'Etat," July 30, 1970 (ANF, 20160129/324).

58. CNEXO, "Coopération franco-américaine dans le domaine de l'océanographie, programme 1973-74," May 29, 1973 (ANF, 20160129/317). Overview based on Jarry, *L'aventure des bathyscaphes*, and on Riffaud and le Pichon, *Expedition FAMOUS*.

59. Heirtzler and Grassle, "Deep-Sea Research."

60. Le Pichon, *Kaiko*.

61. Taira, interview by Martinez-Rius.

62. CNEXO, *Rapport annuel 1973*.

63. CNEXO, *Rapport annuel 1974*.

64. Yves la Prairie, confidential note to Maurice Schumann (minister of scientific tesearch), November 29, 1967 (ANF, 20160129/328).

65. Mero, *Mineral Resources of the Sea*, 127.

66. Stated by the president of Deepsea Venture Inc., John R. Flipse (in photocopied journal clipping, no author, no date, "German Firm Joins Deepsea Ventures' Exploratory Program," enclosed in the report CNEXO, "Projet de création d'une base pour l'exploitation des oceans en Polynesie," 1970 (ANF, 19980125/2).

67. UN General Assembly, First Committee Debate, November 1, 1967.

68. For his book and experience, Mero has been dubbed "the father of ocean mining" (Cruickshank, "John L. Mero: In Memoriam"). His book included economic and commercial estimates of exploiting sands from beaches; gravels and minerals from the continental shelf; minerals dissolved in the seawater; surface sediments (as oozes) from the deep ocean floor; hydrocarbons; and polymetallic nodules (Mero, *Mineral Resources of the Sea*).

69. Mero, *Mineral Resources of the Sea*, 127–241, for a detailed account on manganese nodules' composition, process of formation, and location.

70. Mero started studying seafloor mineral resources at Scripps' Institute of Marine Resources, where he launched the economic study of California offshore phosphorites and manganese (Cruickshank, "John L. Mero: In Memoriam").

71. Pardo affirmed this estimate in his speech at the UN General Assembly. However, Mero gives this estimate only for the cost of processing nodules, not including their exploitation and transportation: 25 US dollars per ton of raw nodules, versus 40 to 100 US dollars of gross commercial value. It is not mentioned on which grounds Mero estimated these costs (Mero, *Mineral Resources of the Sea*, 272).

72. CNEXO, "Synthèse des principales activités océanologiques, du 15 novembre 1969 au 15 janvier 1970" (ANF, 19980125/01).

73. CNEXO, "Synthèse des principales activités océanologiques, du 15 Avril 1970 au 1 Septembre 1970" (ANF, 19980125/2).

74. Yves la Prairie, "Compte-rendu de la mission au Japon, 9–22 Avril 1970," n.d. (ANF, 19980125/02).

75. Georgette Mariani, "Problèmes juridiques poses par l'exploitation du sol et du sous-sol marin: Le rôle du CNEXO," n.d. (ANF, 19980125/02).

76. Alain Sciard (director of international relations), Jean Coulmy (Division of Programs), Jean-Claude Riffaud (Division of Technologies), Lucien Laubier (Groupe scientifique du COB), and Henri-Germain Delauze (president of the offshore tech company COMEX). Yves la Prairie, "Compte-rendu de la mission au Japon, 9–22 Avril 1970," n.d. (ANF, 19980125/02).

77. CNEXO, "Synthèse des principales activités océanologiques, du 15 Avril 1970 au 1 Septembre 1970" (ANF, 19980125/2).

78. Yves la Prairie, "Compte-rendu de la mission au Japon, 9–22 Avril 1970," n.d. (ANF, 19980125/02).

79. La Prairie, "Compte-rendu de la mission au Japon."

80. An industrial group whose 90 percent of the budget came from mining and private companies and 10 percent from the Japanese government.

81. CNEXO, "Synthèse des principales activités océanologiques, du 1 Septembre 1970 au 1 Novembre 1970" (ANF, 19980125/02).

82. CNEXO, *Rapport annuel 1974*.

83. Yves la Prairie, "Note sur l'océanologie au Japon," April 10, 1970 (ANF, 20160129/324).

84. CNEXO, "Projet de création d'une base pour l'exploitation des océans en Polynésie," n.d. (ANF, 19980125/02).

85. For a recently published comprehensive work on the untold consequences of these nuclear tests, based on just-disclosed archival documents, see Philippe and Statius, *Toxique*.

86. CNEXO, "Projet de création d'une base pour l'exploitation des océans en Polynésie," n.d. (ANF, 19980125/02). For more details about the center, see Chesneaux, "Le Centre d'Expérimentation du Pacifique et son impact."

87. Recently, a number of newspaper articles have covered the neglected effects of nuclear radiation during France's nuclear tests in the Pacific. See Henley, "France Has Underestimated Impact."

88. CNEXO, "Projet de création d'une base pour l'exploitation des océans en Polynésie," n.d. (ANF, 19980125/02).

89. Suggested by the American delegation, it came to be known as "the Nixon Amendment." CNEXO, "Projet de création d'une base pour l'exploitation des océans en Polynésie," n.d. (ANF, 19980125/02).

90. CNEXO, *Rapport annuel 1971.*

91. CNEXO, "Projet de création d'une base pour l'exploitation des océans en Polynésie," n.d. (ANF, 19980125/02).

92. Hoffert, *Les nodules polymétalliques.*

93. In Hoffert, *Les nodules polymétalliques,* 135, quoting Yves la Prairie.

94. On the association with the New Caledonian mining company Société Métallurgique Le Nickel—SLN, see CNEXO, *Rapport annuel du groupe de travail CEA-CNEXO: Economie du traitement des nodules polymetalliques* (March–May 1973), May 1973 (ANF, 20080658/40). Half of the company had been acquired in 1974 by the French oil company SNPA, becoming one of its subsidiaries.

95. Retrieved from the dedicated page at the IFREMER website: Henriet, "The Polymetallic Nodules" (https://web.archive.org/web/20210724210216/https://wwz .ifremer.fr/gm_eng/Understanding/Public-Authority-support/Deep-Sea-Mineral -Resources/Polymetallic-nodules).

96. Explained by Michel Hoffert, who was employed in CNEXO's program of nodule exploration (Hoffert, *Les nodules polymétalliques*).

97. Pautot, "TRANSPAC Rapport de mission."

98. Le Suavé et al., "Cadre géologique de concrétions"; Hoffert, *Les nodules polymétalliques.*

99. CNEXO, *Rapport annuel 1974.*

100. CNEXO, "Projet de création d'une base pour l'exploitation des océans en Polynésie," n.d. (ANF, 19980125/02).

101. Hoffert, *Les nodules polymétalliques.*

102. Currently, pushed by a rampant scarcity of strategic minerals, the enthusiasm to exploit deep-sea nodules has revived around the globe. In the last few years, numerous articles have been published in newspapers about the promises and dangers of that new mining industry; see, e.g., Watts, "Race to the Bottom."

103. In Yves la Prairie, "Note pour M. le Chef de Service des Relations Internationales," February 8, 1972 (ANF, 20160129/320).

CHAPTER 6

1. Yvonne Rebeyrol, "Le fond de la Méditerranée est tapissé d'une épaisse couche de sel." *Le Monde*, October 12, 1970.

2. Details from Ryan et al., "Introduction."

3. To expand on the science conducted through scientific ocean drilling, see: Powell, *Mysteries of the Deep.*

4. Hsü, *The Mediterranean Was a Desert*, 36–37.

5. National Academies of Sciences, *High-Performance Bolting Technology*.

6. The Summerland Oil Field was the first productive offshore oil field. The piers extended about 400 meters off the coastline.

7. For a detailed history of the offshore oil industry, see Burleson, *Deep Challenge*. For Shell in particular, see Priest, *Offshore Imperative*.

8. Described in detail in Burleson, *Deep Challenge*, 70–73.

9. Burleson, *Deep Challenge*. Details on the gear, patents, technologies, tests conducted, etc. at 39–45.

10. Historian David K. Van Keuren wrote on the history of the Mohole Project organization and its drilling capabilities in his "Breaking New Ground." A firsthand account of the story, published before the project was canceled, can be read in Bascom, *Hole in the Bottom of the Sea*.

11. Wrigley, "Going Deep."

12. Shor, "Interview with Dr. George and Betty Shor by David K. Van Keuren."

13. Burleson has written the history of Glomar Marine's floating rigs in his *Deep Challenge!*

14. In 1958, the US military had implemented its first global satellite navigation system, *Transit*, which was enhanced in the following decades. In 1967 the US Navy launched a more advanced satellite, Timation, that offered more precise information on location. On the history of navigation systems from the nineteenth century to GPS, see Ceruzzi, *GPS*.

15. All the technical details of the *Glomar Challenger*'s design appear in Burleson, *Deep Challenge!*, and are also described in Van Keuren, "Breaking New Ground."

16. Hsü, *Challenger at Sea*.

17. In scientific ocean drilling, expeditions are identified by number. The DSDP began with Leg 1 and ended in 1983 with Leg 96. The following programs continued the numbering but reset the first digit. The Ocean Drilling Program (ODP) started in 1985 with Leg 100 and ended in 2003 with Leg 210. Beginning with the following Integrated Ocean Drilling Program (unofficially known as "IODP-1"), "Legs" were renamed "Expeditions" and started from 301, ending in 2013 with Expedition 348. The numbering sequence continued through the following phase of scientific ocean drilling, the International Ocean Discovery Program (or "IODP-2"), which covered from Expedition 349 to Expedition 403 in 2024. In 2025, IODP-3—the International Ocean Drilling Programme, read "IODP cubed"—will start with Expedition 501.

18. Historian Tyler Priest has put forward the compelling argument that policies triggered technological innovation for deep-water exploration, rather than the hitherto assumed opposite relation (where technological innovation precedes deep-water exploration). See Priest, "Extraction Not Creation."

19. Lore, "An Exploration and Discovery Model."

20. See Ewing and Antoine, "New Seismic Data."

21. The region was named after the American Commander Charles Dwight Sigsbee, who led the first bathymetric campaigns in the Gulf of Mexico in the 1870s and pioneered the use of wire rope for deep-sea dredging in collaboration with Alexander Agassiz (Sigsbee, *Deep-Sea Sounding and Dredging*.)

22. Ewing et al., "Introduction to Leg 1." Recovering a continuous section means keeping the drilled hole open while bringing a core onboard, and later introducing the drill pipe again in the same hole to recover a second core. The challenge lies in the

difficulty of keeping the hole open during the operation and finding it again when the drill pipe goes in for the second time.

23. Ewing et al., "Introduction to Leg 1"; Burleson, *Deep Challenge!*

24. Yvonne Rebeyrol, "Pour la première fois du pétrole est découvert en mer profonde." *Le Monde*, September 4, 1968.

25. The discovery is mentioned in numerous official reports, including CNEXO, "Synthèse des principales activités océanologiques, du 1r décembre 1968 au 1ᵉʳ février 1969," February 1, 1969 (ANF, 19980125/1).

26. Conseil d'Administration CNEXO, "Procès-verbal de la séance du Conseil d'Administration du CNEXO du 5 décembre 1968," December 5, 1968 (ANF, 19980125/1).

27. CFP, "Sondage d'exploration MISTRAL 1" (AHT, 50ZZ604/32). The next year the well was abandoned as dry, and the company continued exploring the area with geophysical techniques for further sites to drill ("CFP. Documents sur l'exercise 1968" (AHT, *fonds* CFP); "CFP, Documents sur l'exercise 1969" (AHT, *fonds* CFP).

28. Expressed in Xavier le Pichon, "Note à l'attention de DG: Objet: Forages profondes," May 19, 1969 (ANF, 20160129/326).

29. Ryan, *Floor of the Mediterranean Sea*. For more on his biography, see Ryan, "Interview of William Ryan by Tanya Levin."

30. Hsü, *The Mediterranean Was a Desert*, 36.

31. Franco-Russian sedimentologist Wladimir Nesteroff had started his career under the mentorship of Jacques Bourcart. In 1952 he accompanied Jacques Cousteau on the first cruise of the *Calypso* across the Red Sea; later he became a frequent collaborator of researchers at the Lamont Geological Observatory and at Scripps.

32. Maintaining friendly relations was especially important at that moment because la Prairie had just begun his conversations with Edward Wenk Jr. to establish a Franco-American collaboration in ocean sciences. Yves la Prairie, letter to the governor of French Polynesia, March 12, 1969 (ANF, 20160129/326).

33. William B. F. Ryan, interview by Martinez-Rius. Le Pichon invited Ryan to CNEXO when he got his permanent position as scientific advisor to the general director, showing him the facilities, vessels, and national program.

34. Hsü, *The Mediterranean Was a Desert*, 33.

35. All details of the goals of the Mediterranean cruise can be found in Ryan et al., *Initial Reports of the Deep Sea Drilling Project, vol. 13.*

36. Xavier le Pichon, "Note à l'attention de DG: Objet: Forages profondes," May 19, 1969 (ANF, 20160129/326).

37. Hsü, *The Mediterranean Was a Desert* 36–37.

38. Pautot, "Résultats de la campagne de flexo-électro-carottage"; CNEXO, "Contrat IFP—CNEXO: Reconnaissance de structures géologiques en Méditerranée profonde par flexo-électro-carottage," April 16, 1970 (ANF, 20110381/12); CNEXO, "Synthèse des principales activités océanologiques du 15/01/1970 au 15/04/1970," April 26, 1970 (ANF, 19980125/2). Mentioned in DSDP, "Scientific Prospectus for Leg XIII Mediterranean Sea," n.d. (UCSD, b.99 f.7). The usefulness of this campaign for later interpreting the sediments recovered in site 124 is specified in a drafted paper, clipped in a letter from C. Benoit (CNEXO's International Relations Department) to Tom Willey, November 18, 1970 (UCSD, b.88 f.08).

39. DSDP, "Scientific Prospectus for Leg XIII Mediterranean Sea," n.d. (UCSD, b.99 f.7). In his biographical account, Hsü affirmed that seven out of fifteen drill sites were

selected based on CNEXO's and IFP's seismic profiles, but I could not contrast that information with archival sources (Hsü, *The Mediterranean Was a Desert*).

40. Notably a double-pipe system (called a "riser") and a blow-out preventor (BOP).

41. About the Santa Barbara oil spill and its consequences, see Sabol Spezio, *Slick Policy*.

42. Hsü, *The Mediterranean Was a Desert*.

43. As evaporites are accumulated in hypersaline environments, where life is almost impossible, they do not contain microfossils or other elements that enable dating them. Their age can be defined by analyzing the type and species of microfossils that are embedded in under- and overlying sediments. Cita, from the University of Milan, belonged to the first generation of micropaleontologists in Italy and was an expert in studying Italian stratigraphic sections. She had been the first woman to join the *Glomar Challenger* during its second campaign, in 1968, and from then onward she turned her research interests to marine geology. In an interview conducted when she was in her eighties, Cita recalled that mariners did not want a woman onboard, as they believed it would jinx the cruise. Another problem was logistical: cabins were shared. Therefore, the cruise organizers (Americans Melvin Peterson and Terence Edgar) invited a second woman, the Canadian paleontologist Catherine Nigrini. Both Cita and Nigrini worked as scientists of the onboard party, together with other five male scientists, during this DSDP's second expedition. (Cita, interview by Dirk Simon).).

44. According to the scientific party onboard the 1970s cruise, the refilling of the Mediterranean basin happened in less than a thousand years (Ryan et al., "Origin of the Mediterranean Evaporites"). This megaflooding has been named the Zanclean megaflood (see García-Castellanos et al., "Zanclean Megaflood").

45. The name came from Italy, as Italian geologists had previously identified similar salt deposits along Sicily. The time of deposition had already been named "Messinian."

46. Apparently, French geologist Wladimir Nesteroff put forward the shallow basin hypothesis, which received large support within France's academic circles. The idea of a shallow, totally dried Mediterranean stemmed from the hypothesis Jacques Bourcart (Nesteroff's mentor) had put forward, but this idea was soon discarded due to evidence showing that the Mediterranean was a deep marine basin.

47. Letter from Melvin N. A. Peterson (chief scientist DSDP) to Yves la Prairie, September 21, 1970 (UCSD, b.99 f.8).

48. Yvonne Rebeyrol, "Le fond de la Méditerranée est tapissé d'une épaisse couche de sel." *Le Monde*, October 12, 1970"; Hsü, "When the Mediterranean Dried Up."

49. CNEXO, "Contrat IFP-CNEXO n. 71/326," November 4, 1971 (ANF 20110381/14).

50. Until the DSDP expedition in the Mediterranean took place, in the mid-1970s, there were only two exploration and exploration leases in the western Mediterranean, both in the continental shelf: CFP's "Permis du Golfe du Lion" and "Permis Corse Maritime." Data based on the map: Direction des Carburantes, "Périmètres des titres miniers d'hydrocarbures," January 1, 1973 (ANF, 19980125/49).

51. This increasing interest in plate tectonics and the Messinian Salinity Crisis is well reflected in the biannual volumes of the CIESM (e.g., CIESM, *Rapports et procès-verbaux des réunions*, 22). I have analyzed how the discovery of deep salt domes and the new framework of plate tectonics prompted collaboration and exchanges between the oil industry and academic geologists in Martinez-Rius, "An Open Secret."

52. Thompson, "Plate Tectonics"; Bullard, "Overview of Plate Tectonics"; Fischer, "Epilogue." On the impact of the theory on the US oil industry, see Edmundson, "Tectonic Shocks in the Oil Industry."

53. Aubouin, *Geosynclines.*

54. Oil companies expressed their interest in exploring the Mediterranean and its salt layer in internal reports; see, e.g., CFP, "Programme forage profond," June 30, 1971 (ANF, 19980125/25); CEPM, "L'exploration des hydrocarbures en France: Résultats récents et nouvelles approaches," n.d. (ANF, 1998125/25); CNEXO, "Contrat IFP-CNEXO n. 71/326," November 4, 1971 (ANF, 20110381/14).

55. CNEXO, "Procès-verbal de la séance du Comité Scientifique et Technique du 28 novembre 1973," November 28, 1973 (ANF, 19980125/4).

56. Twenty-five geologists, engineers, and technicians from European and American companies attended, including Agip, Apex, Auximi Petroleos, BP Italiana, British Petroleum, CFP France and Italy, CFP, Coparex Española, Deutsche Texaco, Elf Italiana Mineraria, Gulf Oil Italia, IFP, Shell, SFP, SNPA and Techneco. See CIESM, *Rapports et procès-verbaux des réunions, 22.*

57. Hsü and Ryan, "Report on Deep Sea Drilling Project"; Nesteroff, "La crise de salinité méssinienne"; Edgar et al., "Plans for Future Drilling."

58. A note on geology: How could the evaporites under the seabed have been eroded? This can only be explained by assuming that, in the past, they were exposed to atmospheric conditions such as wind and rain, which would have worn them down.

59. Besides Hsü (KTH, Switzerland) and Ryan (Lamont Geological Observatory, US), the members of the DSDP Mediterranean Advisory Panel were Enrico Bonatti (Rosenstiel School of Marine and Atmospheric Science, US), David Ross (Woods Hole Oceanographic Institution), Maria Bianca Cita (University of Milan, Italy), Mikhail V. Muratov (P. P. Shirshov Institute of Oceanology, USSR), Frank H. Fabricius (Technische Universität, Munich, Germany), and Lucien Montadert (IFP, France).

60. See, e.g., Montadert et al., "De l'âge tertiaire de la série salifère."

61. Montadert, interview by Martinez-Rius.

62. After the IFP's multichannel seismic surveys, the French oil companies followed (through CEPM), together with CNEXO and foreign research institutions (like the Istituto Nazionale di Oceanografia e di Geofisica Sperimentale [OGS], Italy's National Institute of Oceanography and Applied Geophysics, in Trieste). In Biju-Duval et al., "Structure and Evolution."

63. DSDP, "Minutes of the Mediterranean Panel Meeting," January 17, 1974 (UCSD b.8 f.13); Hsü and Montadert, "DSDP Scientific Prospectus for Leg 42, Part A—Mediterranean Sea," March 5, 1975 (UCSD b.104 f.14); Hsü, Montadert et al., "Introduction and Explanatory Notes."

64. Hsü and Montadert, "DSDP Scientific Prospectus for Leg 42, Part A—Mediterranean Sea," March 5, 1975 (UCSD b.104 f.14). The others were retained based on data from the Italian OGS, the German campaign with the *Meteor*, and British cruises.

65. Terence Edgar, letter to Bettye [no surname given], "Leg 39 Mediterranean," October 3, 1973 (UCSD, b.104 f.15).

66. Initially, targeted nations were the US, Canada, the UK, Japan, and the Netherlands; but it later extended to Portugal, Rhodesia, and South Africa.

67. In 1973, the US consumed 17 million barrels of oil per day. Of those, 2,095 came from OPEC countries. Data from US Energy Information Administration: https://www.eia.gov/dnav/pet/hist/LeafHandler.ashx?n=PET&s=MCRIMXX2&f=A.

68. Pratt et al., *Offshore Pioneers*; see also Yergin, *The Prize*.

69. The D/V *Pélican* was designed by the Dutch firm GustoMSC and built in the shipyards of IHC, Rotterdam. CFP, "Programme de forage profond," June 30, 1971 (ANF, 19980125/25).

70. Direction des Hydrocarbures, "Offshore profond," March 3, 1980 (ANF, 19980125/49). The CFP considered sharing the research burden of "nonindustrial" campaigns with CNEXO and CEPM, as it had been doing with seismic surveys in hybrid campaigns (Harbonn, "Projet: Forages sous grande immersion CFP," June 24, 1971 [ANF, 19980125/25]). In 1972, the CFP drilled the borehole AUTAN in the Gulf of Lion (CFP, *Rapport annuel 1972*).

71. The CFP was closely following technological innovations introduced in the *Glomar Challenger* and comparing them with the *Pélican* drilling capabilities: CFP, "Forage sus grande immersion," May 16, 1972 (ANF, 19980125/49).

72. The boreholes drilled during the DSDP Leg 42 fell outside the French oil industry's leases in the deep Mediterranean (Direction des Carburantes, "Périmètres des titres miniers d'hydrocarbures," January 1, 1973 [ANF, 19980125/49]).

73. CEPM, "Possibilités pétrolières des zones marines profondes de la Méditerranée occidentale," 1979 (ANF, 19980125/36).

74. CFP, "Bilan des activités pétrolières marines en Espagne et en Italie," n.d. (ANF, 19980125/25). ENIEPSA, Empresa Nacional de Investigaciones y Explotaciones Petrolíferas S.A., was a Spanish state-owned company under the National Institute of Hydrocarbons. Particularly active during the 1980s both on land and offshore, in 1987 it was merged into Repsol, the newly created state energy company that brought together various Spanish oil and gas enterprises.

75. CEPM, "L'exploration des hydrocarbures en France: Résultats récents et nouvelles approches," n.d. (ANF, 1998125/25).

76. Gaston-Breton, *Total, un esprit pionnier*. The second oil crisis started in the aftermath of the Iranian revolution, caused by strikes of workers in oil-producing facilities and the subsequent Iraq-Iran war. This provoked a drop-down of oil production that deeply affected Western countries, since Iran was the second largest exporter of oil. See Yergin, *The Prize*.

77. On the 1980s decadence of the Gulf of Mexico, see Priest, "Extraction Not Creation." On the North Sea boom of the 1980s, see Pratt et al., *Offshore Pioneers*.

78. In planning future cruises, the American organizers prioritized less explored areas or those more geologically interesting from a global tectonic perspective (the Mediterranean basin had two cruises within a decade, sufficient if it is considered that the DSDP aimed at touring all the world's oceans). The *JOIDES Resolution* did come to the Mediterranean in 1987, but only to study the Tyrrhenian Sea (during the Ocean Drilling Program Leg 107).

EPILOGUE

1. Yvonne Rebeyrol, "Au colloque de l'ASTEO le réalisme remplace l'illusion lyrique' dans les discours officiels sur l'océanologie." *Le Monde*, January 25, 1982.

2. Crochet, *Marées noires*.

3. "Dès janvier 1978 un rapport officiel à usage interne dénonçait les insuffisances des moyens de lutte contre la pollution." *Le Monde*, March 29, 1978.

4. Paccalet, *Jacques-Yves Cousteau*.

5. Cousteau and Jacquier, *Français, on a volé ta mer*, 35.

6. Cousteau and Jacquier, *Français, on a volé ta mer*, 19.

7. Cousteau and Jacquier, *Français, on a volé ta mer*, 19.

8. Oreskes, *Science on a Mission*.

9. Henri Jacquier described Cousteau's disagreements with Yves la Prairie as well as his tense relation with CNEXO in Jacquier, "L'aventure de Jacques-Yves Cousteau."

10. Jacquier, "L'aventure de Jacques-Yves Cousteau."

11. Naomi Oreskes has pointed to similar affirmations among American oceanographers who witnessed firsthand the growth of oceanography in their country during the Cold War. She offers an in-depth analysis behind this affirmation in her epilogue to *Science on a Mission*.

12. Lacour, *Terre d'innovations*.

13. Doel et al., "Extending Modern Cartography"; Weir, *An Ocean in Common*; Rainger, "Science at the Crossroads"; Hamblin, *Oceanographers and the Cold War*; Oreskes, "Context of Motivation."

14. Oreskes, "Context of Motivation."

15. See its history in Chatry, *Il était une fois l'IFREMER*.

16. The principles and initiatives of the United Nations Decade of Ocean Science for Sustainable Development are particularly representative: see UNESCO-IOC, *United Nations Decade*.

17. Sénat, "Rapport d'Information."

18. Note that the program's name reflects this shift in leadership: instead of using the American spelling for "program," IODP-3 has resorted to the British spelling "programme." The new IODP-3 will run the Japanese scientific drillship *Chikyu* and the ECORD-provided mission-specific platforms.

19. Argument based on the works and ideas of Cronon, *Nature's Metropolis*; Rozwadowski, *Vast Expanses*.

Bibliography

ARCHIVAL SOURCES

Archives Nationales de France (ANF), Pierrefitte-sur-Seine, Paris
Archives Diplomatiques de France (ADF), La Courneuve, Paris
Archives du Muséum National d'Histoire Naturelle (MNHN), Paris
Archives Historiques du Groupe TOTAL (AHT), Paris
Service Historique de la Défense (SHD), Vincennes and Châllerault, France
NATO Archives, NATO, https://archives.nato.int/nato-archives-online
University of California San Diego, Special Collections and Archives (UCSD), La Jolla,
    California

ORAL HISTORY INTERVIEWS

Boillot, Gilbert. Interview with Gilbert Boillot by Beatriz Martinez-Rius. Nice, France.
    May 31, 2021 (unpublished).
Cita, Maria Bianca. Interview with Maria Bianca Cita by Dirk Simon. Brisighella, Sicily.
    2012 (unpublished).
Lalou, Claude. "Interview with Claude Lalou by Jean-François Piccard." Archives
    Orales du CNRS. July 19, 1986.
Montadert, Lucien. Interview with Lucien Montadert by Beatriz Martinez-Rius. Rueil-
    Malmaison, France. March 13, 2019 (unpublished).
Ryan, William. "Interview of William Ryan by Tanya Levin." Niels Bohr
    Library and Archives, American Institute of Physics. July 1, 1998.
    www.aip.org/history-programs/niels-bohr-library/oral-histories/22894.
———. Interview with William B. F. Ryan by Beatriz Martinez-Rius. Virtual
    interview at Columbia University, New York City, from Paris. November 15, 2019
    (unpublished).
Shor, George, and Betty Shor. "An Interview with Dr. George and Betty Shor by David
    K. Van Keuren." University of California San Diego, Special Collections, SIO Oral
    Histories. March 19, 1995. https://library.ucsd.edu/speccoll/siooralhistories/
    Shor1995.pdf.
Taira, Asahiko. Interview with Asahiko Taira by Beatriz Martinez-Rius. Tokyo.
    December 20, 2023. https://talesofoceanscience.com/asahiko-taira-part-2/.

PUBLISHED SOURCES

Abulafia, David. *The Great Sea: A Human History of the Mediterranean.* Oxford University Press, 2011.

Adamson, Matthew. "Les Liaisons Dangereuses: Resource Surveillance, Uranium Diplomacy and Secret French-American Collaboration in 1950s Morocco." *British Journal for the History of Science* 49, no. 1 (March 2016): 79–105.

Adler, Antony. "Cold War Science on the Seafloor." In *Neptune's Laboratory: Fantasy, Fear, and Science at Sea.* Harvard University Press, 2019.

———. "Deep Horizons: Canada's Underwater Habitat Program and Vertical Dimensions of Marine Sovereignty." *Centaurus* 62, no. 4 (2020): 763–82.

———. "Marine Science for the Nation or for the World?" In *Neptune's Laboratory: Fantasy, Fear, and Science at Sea.* Harvard University Press, 2019.

Alinat, Jean, Günter Giermann, and Oliver Leenhardt. "Reconnaissance sismique des accidents de terrain en mer Ligure." *C. R. Académie des sciences de Paris* B 262 (1966): 1311–14.

Aubouin, Jean. *Geosynclines.* Elsevier, 1965.

Ballard, Robert D. "The History of Woods Hole's Deep Submergence Program." In *50 Years of Ocean Discovery: National Science Foundation 1950–2000,* ed. Ocean Studies Board and National Research Council. National Academies Press, 2000.

Barkenbus, Jack N. "The Politics of Ocean Resource Exploitation." *International Studies Quarterly* 21, no. 4 (Special Issue on International Politics of Scarcity) (December, 1977): 675–700.

Bascom, Willard. *A Hole in the Bottom of the Sea: The Story of the Mohole Project.* Doubleday, 1961.

Bates, Charles C., Thomas F. Gaskell, and Robert B. Rice. *Geophysics in the Affairs of Man.* Pergamon Press, 1982.

Beltran, Alain. "L'industrie pétrolière en France pendant la première guerre mondiale: une prise de conscience tardive." In *L'industrie dans la grande guerre: Colloque des 15 et 16 novembre 2016, edited by Patrick Friedenson and Pascal Griset.* Institut de la gestion publique et du développement économique, Comité pour l'histoire économique et financière de la France, 2018.

Beuzart, Paul. *Nestlante II.* Centre National pour l'Exploitation des Océans, 1981.

Biju-Duval, Bernard, Jean Letouzey, and Lucien Montadert. "Structure and Evolution of the Mediterranean Basins." In *Initial Reports of the Deep Sea Drilling Project* 42, pt. 1, edited by Kenneth J. Hsü, Lucien Montadert, et al. US Government Printing Office, 1978.

Boillot, Gilbert. *Comment l'idée vient au géologue.* Orizons, 2021.

———. "Des marges continentales atlantiques aux chaînes plissées: Les recherches françaises entre 1967 et 2000." *Travaux du comité français d'histoire de la géologie* 3, vol. 26, no. 1 (2012): 1–24.

———. "1956–2012: La géologie marine à l'Observatoire Océanologique de Villefranche-sur-Mer." *Travaux du Comité Français d'Histoire de la Géologie* 33 (2019).

Bonneuil, Christophe, and Jean-Baptiste Fressoz. *The Shock of the Anthropocene: The Earth, History, and Us.* Verso, 2016.

Bourcart, Jacques. "Colloque de géologie sous-marine." *Comité Central d'Océanographie et d'Étude des Côtes. Bulletin d'information* 8, no. 7 (October 1955): 331–70.

———. "Essai de classement des formations continentales quaternaires du Maroc occidental." *Comptes rendus séances Société Géologique de France* (1931): 256–59.

———. *Géographie du fond des mers.* Éditions Payot, 1949.

———. "Géologie sous-marine de la baie de Villefranche." *Annales de l'Institut Océanographique* 3, no. 33 (1957): 137–200.

———. "Hypothèses sur la genèse des gorges sous-marines." *Comité Central d'Océanographie et d'Étude des Côtes. Bulletin d'information* 2, no. 9 (November 1950): 21–39.

———. *La connaissance des profondeurs océaniques.* Société d'Édition d'Enseignement Supérieur, 1964.

———. "La marge continentale: Essai sur les régressions et les transgressions marines." *Bulletin de la Société Géologique de France* 8, no. 4 (1938): 393–474.

———. *Le fond des océans.* Presses Universitaires de France, 1954.

———. "Le 'Rech' Lacaze-Duthiers, cañon sous-marin du plateau continental du Roussillon." *Comptes rendus de l'Académie des Sciences* 226 (n.d.): 1632–33.

———. *Les confins albanais administrés par la France, 1916–1920: Contribution à la géographie et à la géologie de l'Albanie moyenne.* Librairie Delagrave, 1922.

———. "Les dépôts du second cycle miocène du Maroc occidental." *Comptes rendus hebdomadaires des séances de l'Académie des Sciences* 195 (October 24, 1932): 736.

———. "Premiers résultats d'une étude du quaternaire du Maroc." *Bulletin de la Société Géologique de France* 27, no. 4 (1927): 3–33.

———. *Problèmes de géologie sous-marine: Le précontinent—Le littoral et sa protection—La stratigraphie sous-marine.* Masson et Cie Éditeurs, 1958.

———. "Topographie sous-marine et sédimentation actuelle." In *Comptes rendus de la dix-neuvième session alger 1952*, by Congrès Géologique International, vol. 4. Imprimerie Protat, 1953.

Bourée, Henri. *L'océanographie vulgarisée.* C. Delagrave, 1913.

Bourgoin, Jean. "Henri Lacombe, 1913–2000." *La jaune et la rouge* (February 2001): 51–52.

Bowker, Geoffrey C. *Science on the Run: Information Management and Industrial Geophysics at Schlumberger, 1920–1940.* MIT Press, 1994.

Braudel, Fernand. *La Méditerranée et le monde méditerranéen à l'epoque de Philippe II.* Librarie A. Colin, 1949.

Briseid, E., and Jean Mascle. "Structure de la marge continentale norvégienne au débouché de la mer de Barentz." *Marine Geophysical Researches* 2 (1975): 231–41.

Brunn, Anton F. "The International Advisory Committee on Marine Sciences of UNESCO." *AIBS Bulletin* 8, no. 1 (January 1958): 12–14.

Bullard, Edward. "Overview of Plate Tectonics." In *Petroleum and Global Tectonics*, edited by Alfred G. Fischer. Princeton University Press, 1975.

———. "William Maurice Ewing." *Biographical Memoirs of Fellows of the Royal Society* 21 (November 1975).

Burleson, Clyde W. *Deep Challenge! The True Epic Story of Our Quest for Energy Beneath the Sea.* Gulf, 1999.

Bush, Vannevar. *Science, the Endless Frontier.* US Government Printing Office, 1945.

Buzan, Barry. *Seabed Politics.* Praeger, 1976.

Camprubí, Lino. "'No Longer an American Lake': Depth and Geopolitics in the Mediterranean." *Diplomatic History* 44, no. 3 (June 2020): 428–46.

———. "Resource Geopolitics: Cold War Technologies, Global Fertilizers, and the Fate of Western Sahara." *Technology and Culture* 56, no. 3 (2015): 676–703.

———. "The Sonic Construction of the Ocean as the Navy's *Operating Environment*." In *Navigating Noise*, edited by Nathanja van Dijk, Kerstin Ergenzinger, Christian Kassung, and Sebastian Schwesinger. Verlag der Buchhandlung Walther König, 2017.

Camprubí, Lino, and Philipp Lehmann. "The Scales of Experience: Introduction to the Special Issue 'Experiencing the Global Environment.'" *Studies in History and Philosophy of Science* 70 (August 2018): 1–5.

Camprubí, Lino, and Sam Robinson. "A Gateway to Ocean Circulation: Surveillance and Sovereignty at Gibraltar." *Historical Studies in the Natural Sciences* 46, no. 4 (2016): 429–59.

Cantoni, Roberto. *Oil Exploration, Diplomacy, and Security in the Early Cold War: The Enemy Underground*. Routledge, 2017.

Cassand, Jean, Jean-Pierre Fail, and Lucien Montadert. "Sismique réflexion en eau profonde (Flexotir)." *Geophysical Prospecting* 18, no. 4 (December 1970): 600–614.

Caulfield, David D. "Predicting Sonic Pulse Shapes of Underwater Spark Discharges." *Deep-Sea Research*, no. 9 (1962): 339–48.

Ceruzzi, Paul E. *GPS*. MIT Press, 2018.

Chapelle, Jean. "Le pétrole et la mer." *Sondages: Bulletin interieur de l'IFP*, no. 42. IFP (1966): 3–14.

Chatry, Gilles. *Il était un fois l'IFREMER*. Éditions Quae, 2021.

Cherruau, François, and Archille Ferrari. "André Giraud." *La jaune et la rouge* 529 (November 1997): 38–40.

Chesneaux, Jean. "Dispositif émetteur d'ondes sonores." Office de la Propriété Intellectuelle du Canada, CA 723001-A (07/12/1965).

———. "Le Centre d'Expérimentation du Pacifique et son impact." In *Tahiti après la bombe: Quel avenir pour la Polynésie?*, edited by Jean Chesneaux. L'Harmattan, 1995.

Churchill, Robin R., and Alan Vaughan Lowe. *The Law of the Sea*. 2nd ed. Manchester University Press, 1988.

Commission Internationale pour l'Exploration Scientifique de la Mer Méditerranée (CIESM). *Rapports et procès-verbaux des réunions*, 22. CIESM, 1973.

Clarke, Arthur C. *The Deep Range. Bantam Books, 1991*.

———. *Dolphin Island: A Story of the People of the Sea*. Holt, Rinehart and Winston, 1964.

CNEXO. *Rapport annuel 1970*. Centre National pour l'Exploitation des Océans, 1971.

———. *Rapport annuel 1971*. Centre National pour l'Exploitation des Océans, 1972.

———. *Rapport annuel 1972*. Centre National pour l'Exploitation des Océans, 1973.

———. *Rapport annuel 1973*. Centre National pour l'Exploitation des Océans, 1974.

———. *Rapport annuel 1974*. Centre National pour l'Exploitation des Océans, 1975.

Collis, Christy, and Klaus Dodds. "Assault on the Unknown: The Historical and Political Geographies of the International Geophysical Year (1957–8)," *Journal of Historical Geography* 34, no. 4 (2008): 555–73.

Colloques internationaux du CNRS. *La topographie et la géologie des profondeurs océaniques: Nice, 5–12 Mai 1959*. CNRS, 1959.

Colloques nationaux du CNRS. *Océanographie géologique et géophysique de la Méditerranée occidentale: Villefranche sur Mer, 4 au 8 Avril 1961*. Éditions du Centre National de la Recherche Scientifique, 1962.

Combaz, André. "Le comité d'études pétrolières marines." *Annales de géographie* 88, no. 486 (1979): 145–49.

————. "Les premières découvertes de pétrole au Sahara dans les années 1950: Le témoignage d'un acteur." *Travaux du Comité Français d'Histoire de la Géologie* 3, 16 (June 19, 2002).

Conseil économique et social. "Étude sur la recherche océanographique et ses applications." In *Avis et rapports* 18 (1962): 737–47.

Cousteau, Jacques-Yves, and Henri Jacquier. *Français, on a volé ta mer*. Robert Lafont, 1982.

Crochet, Bernard. *Marées noires: 50 ans de catastrophes écologiques*. Éditions Ouest France, 2018.

Cronon, William. *Nature's Metropolis: Chicago and the Great West*. Norton, 1992.

Cruickshank, Michael J. "John L. Mero: In Memoriam." *Marine Georesources and Geotechnology* 20 (2002): 85–86.

Curray, Joseph R. "*Francis P. Shepard, 1897–1985*." In *Coming of Age: Scripps Institution of Oceanography: A Centennial Volume, 1903–2003*, edited by Robert L. Fisher, Edward D. Goldberg, and Charles S. Cox. Scripps Institution of Oceanography, University of California, San Diego, 2003.

Debyser, Jacques, Xavier le Pichon, and Lucien Montadert, eds. *Histoire structurale du Golfe de Gascogne*. Éditions Technip, 1971.

Delacour, Jacques, and André Castela. "Appareillage de forage sous-marin." Brevet d'invention no. 1.482.823. Ministère de l'Industrie, Service de la Propriété Industrielle, July 2, 1965.

Delacour, Jacques, Jean Parola, and Pierre Grolet. "Dispositif pour le carottage sous-marine." Office de la Propriété Intellectuelle du Canada, Patent CA 726977A. February 1, 1966.

Délégation Général de la Recherche Scientifique et Technique (DGRST). *Les actions concertées: Rapport d'activité 1961*. DGRST, 1962.

Dietz, Robert S. "Continent and Ocean Basin Evolution by Spreading of the Sea Floor." *Nature* 190 (June 1960): 854–57.

Doel, Ronald E., and Kristine C. Harper. "Prometheus Unleashed: Science as a Diplomatic Weapon in the Lyndon B. Johnson Administration." *Osiris* 21, no. 1 (January 2006): 66–85.

Doel, Ronald E., Tanya J. Levin, and Mason K. Marker. "Extending Modern Cartography to the Ocean Depths: Military Patronage, Cold War Priorities, and the Heezen–Tharp Mapping Project, 1952–1959." *Journal of Historical Geography* 32, no. 3 (2006): 605–26.

Dolan, John R. "An Early Example of International Oceanography: The Last Voyage of the *Carnegie* (1928–1929)." *Bulletin of Limnology and Oceanography* 30, no. 3 (2001): 92–96.

Edgar, Terence, William B. F. Ryan, and Kenneth J. Hsü. "Plans for Future Drilling in the Mediterranean and Black Sea." In *Rapports et procès-verbaux des réunions* 22, edited by Commission Internationale pour l'Exploration Scientifique de la Mer Méditerranée (CIESM). CIESM, 1973.

Edgerton, David. "'The Linear Model' Did Not Exist: Reflections on the History and Historiography of Science and Research in Industry in the Twentieth Century." In *The Science-Industry Nexus: History, Policy, Implications*, edited by Karl Grandin, Nina Wormbs, and Sven Widmalm. Science History Publications, 2004.

———. *The Rise and Fall of the British Nation: A Twentieth-Century History*. Penguin Books, 2018.

Edgerton, Harold E. "The 'Boomer' Sonar Source for Seismic Profiling." *Journal of Geophysical Research* 69, no. 14 (July 15, 1964): 3033–42.

Edmundson, Mark. "Tectonic Shocks in the Oil Industry." *AAPG Explorer*, January 11, 2016.

Elden, Stuart. "Land, Terrain, Territory." *Progress in Human Geography* 34, no. 6 (December 2010): 799–817.

Esestime, Paolo, Ashleigh Hewitt, and Neil Hodgson. "Zohr—A Newborn Carbonate Play in the Levantine Basin, East-Mediterranean." *First Break* 34, no. 2 (February 1, 2016): 87–93.

Ewing, Maurice, and John Antoine. "New Seismic Data Concerning Sediments and Diapiric Structures in Sigsbee Deep and Upper Continental Slope, Gulf of Mexico," *AAPG Bulletin* 50, no. 3 (March 1966): 479–504.

Ewing, Maurice, and Frank Press. "Geophysical Contrasts Between Continents and Ocean Basins." Special paper, *Geological Society of America* 62 (January 1, 1955).

Ewing, Maurice, J. Lamar Worzel, and Creighton A. Burk, "Introduction to Leg 1." In *Initial Reports of the Deep Sea Drilling Project* 1, edited by Maurice Ewing and J. Lamar Worzel, 3–9. US Government Printing Office, 1969.

Fahlquist, Davis Armstrong. *"Seismic Refraction Measurements in the Western Mediterranean Sea."* PhD diss., Massachusetts Institute of Technology, 1963.

Fail, Jean Pierre, Lucien Montadert, Jean Raymond Delteil, Pierre Valéry, Philippe Patriat, and Roland Schlich. "Prolongation des zones de fractures de l'océan atlantique dans le Golfe de Guinée." *Earth and Planetary Science Letters* 7 (1970): 413–19.

Felt, Hali. *Soundings: The Story of the Remarkable Woman Who Mapped the Ocean Floor*. Henry Holt, 2012.

Finley, Carmel. *All the Boats on the Ocean*. University of Chicago Press, 2017.

Fischer, Alfred G., ed. "Epilogue." In *Petroleum and Global Tectonics*, 321–23. Princeton University Press, 1975.

Frankel, Henry R. *The Continental Drift Controversy*. Cambridge University Press, 2012.

Friedman, Alan G., and Cynthia A. Williams. "The Group of 77 at the United Nations: An Emergent Force in the Law of the Sea." *San Diego Law Review* 16 (1979): 555–74.

García-Castellanos, Daniel, Aaron Micallef, Ferran Estrada, et al. "The Zanclean Megaflood of the Mediterranean—Searching for Independent Evidence." *Earth-Science Reviews* 201 (2020).

García-Castellanos, Daniel, and Ricarda Vidal. "Alternative Mediterraneans Six Million Years Ago: A Model for the Future?" In *Alternative Worlds*, edited by Ricarda Vidal and Ingo Cornils. Peter Lang, 2015.

Gärdebo, Johan. *"Environing Technology: Swedish Satellite Remote Sensing in the Making of Environment 1969–2001."* PhD diss., Royal Institute of Technology, 2019.

Gaskell, Thomas F., and John C. Swallow. "Seismic Refraction Experiments in the Indian Ocean and in the Mediterranean Sea." *Nature* 172 (September 19, 1953): 535–37.

Gaston-Breton, Tristan. *Total, un esprit pionnier*. Pollina Éditions Textuel, 2019.

Gavouneli, Maria. "From Uniformity to Fragmentation: The Ability of the UN Convention on the Law of the Sea to Accommodate New Uses and Challenges." In *Unresolved Issues and New Challenges to the Law of the Sea*, edited by Anastasia Strati, Maria Gavouneli, and Nikos Skourtos. Martinus Nijhoff, 2006.

Gennesseaux, Maurice, and Jean Mascle. "La naissance et le développement de la géologie marine à Villefranche-sur-Mer: Des années 1950 au milieu des années 1980." *Travaux du Comité Français d'Histoire de la Géologie (COFRHIGÉO)* 3rd ser., vol. 26, no. 10 (2012): 193–233.

Giraud, André, and Xavier Boy de la Tour. *Géopolitique du pétrole et du gaz*. Éditions Technip, 1987.

Gorove, Stephen. "The Concept of 'Common Heritage of Mankind': A Political Moral or Legal Innovation?" *San Diego Law Review* 9, no. 3 (May 1, 1972): 390.

Gougenheim, André. "Funérailles de Jacques Bourcart." *Académie des Sciences: Notices et Discours* 5 (June 28, 1965): 246–51.

———. "Les canyons sous-marins de la côte sud de France." *Comité Central d'Océanographie et d'Étude des Côtes. Bulletin d'information* 2, no. 3 (March 1950): 93–95.

Graham, John R., and John A. Reed. "Glomar Challenger Deep Sea Drilling Vessel." *Journal of Petroleum Technology* 21 (1969): 1263–74.

Groupe scientifique du COB. *Résultats des campagnes du N.O. Jean CHARCOT: Campagne Noratlante: 3 août-2 novembre 1969*. Publications du Centre National pour l'Exploitation des Océans (CNEXO), serie: Résultats des campagnes à la mer, no. 1–1971, 1971. https://archimer.ifremer.fr/doc/00413/52412/.

Guilcher, André. "Francis P. Shepard (1897–1985), père de la géomorphologie marine." *Annales de Géographie* 95, no. 527 (1986): 87–98.

Hamblin, Jacob Darwin. *Oceanographers and the Cold War Disciples of Marine Science*. University of Washington Press, 2005.

Hannigan, John A. *The Geopolitics of Deep Oceans*. Polity, 2015.

Hardin, Garrett. "The Tragedy of the Commons." *Science* 162, no. 3859 (1968): 1243–48.

Hardy, Noah. "Letter from Paris: Exploring the Shelf." *Science* 95, no. 15 (April 12, 1969): 366.

Hartingh, France de. "La position française à l'égard de la convention de Genève sur le plateau continental." In *L'Annuaire français de droit international* 11 (1965): 725–34.

Harvie, Christopher T. *Fool's Gold: The Story of North Sea Oil*. Penguin Books, 1995.

Hecht, Gabrielle. *The Radiance of France: Nuclear Power and National Identity After World War II*. MIT Press, 1998.

Heezen, Bruce. "The Rift in the Ocean Floor." *Scientific American* 203, no. 4 (1960): 98–110.

Heezen, Bruce C., Marie Tharp, and Maurice Ewing. "The Floors of the Oceans: 1. The North Atlantic." *Special paper, Geological Society of America*, 1959.

Heirtzler, J. R., and J. F. Grassle. "Deep-Sea Research by Manned Submersibles." *Science* 194, no. 4262 (October 15, 1976): 294–99.

Heirtzler, James R., Xavier le Pichon, and Gregory J. Baron. "Magnetic Anomalies over the Reykjanes Ridge." *Deep-Sea Research* 13, no. 3 (1996): 427–32.

Henley, Jon. "France Has Underestimated Impact of Nuclear Tests in French Polynesia, Research Finds." *The Guardian*, March 9, 2021. https://www.theguardian.com/world/2021/mar/09/france-has-underestimated-impact-of-nuclear-tests-in-french-polynesia-research-finds.

Henriet, Jean-Pierre. "The Polymetallic Nodules." April 19, 2016. https://web.archive.org/web/20210724210216/https://wwz.ifremer.fr/gm_eng/Understanding/Public-Authority-support/Deep-Sea-Mineral-Resources/Polymetallic-nodules.

Hersey, John Brackett. "Continuous Reflection Profiling." In *The Sea: Ideas and Observations on Progress in the Study of the Seas*, vol. 3, edited by Maurice Neville Hill. John Wiley and Sons, 1963.

———. "Sedimentary Basins of the Mediterranean Sea." *Woods Hole Oceanographic Institution Contributions* 1628 (1965): 75–91.

Hoffert, Michel. *Les nodules polymétalliques dans les grands fonds océaniques: Une extraordinaire aventure minière et scientifique sous-marine.* Société Géologique de France Vuibert, 2008.

Höhler, Sabine. "Depth Records and Ocean Volumes: Ocean Profiling by Sounding Technology, 1850–1930." *History and Technology* 18, no. 2 (2002): 119–54.

———. "A Sound Survey: The Technological Perception of Ocean Depth, 1550– 1930." In *Transforming Spaces: The Topological Turn in Technology Studies*, edited by Mikael Hård, Andreas Lösch, and Dirk Verdicchio. Online publication of the International Conference held in Darmstadt, Germany, March 22–24, 2002 (2003).

Hollick, Ann L. *U.S. Foreign Policy and the Law of the Sea.* Princeton University Press, 1981.

Hollick, Ann L., and Robert E. Osgood. *New Era of Ocean Politics.* Johns Hopkins University Press, 1975.

Hsü, Kenneth J. *Challenger at Sea.* Princeton University Press, 1992.

———. *The Mediterranean Was a Desert: A Voyage of the Glomar Challenger.* Princeton University Press, 1987.

———. "When the Mediterranean Dried Up." *Scientific American* 227, no. 6 (1972): 26–36.

Hsü, Kenneth J., and William B. F. Ryan. "Report on Deep Sea Drilling Project in the Mediterranean." In *Rapports et procès-verbaux des réunions* 22, edited by Commission Internationale pour l'Exploration Scientifique de la Mer Méditerranée (CIESM),. CIESM, 1973.

Hsü, Kenneth J., Lucien Montadert, et al. "Introduction and explanatory notes." In *Initial Reports of the Deep Sea Drilling Project*, edited by Kenneth J. Hsü, Lucien Montadert, and the science party of DSDP Leg 42. US Government Printing Office, 1978.

Institut Français du Pétrole. *Rapport annuel de l'Institut Français du Pétrole 1963.* IFP, 1964.

———. *Rapport annuel de l'Institut Français du Pétrole 1964.* IFP, 1965.

———. *Rapport annuel de l'Institut Français du Pétrole 1965.* IFP, 1966.

———. *Rapport annuel de l'Institut Français du Pétrole 1967.* IFP, 1968.

———. *Rapport annuel de l'Institut Français du Pétrole 1968.* IFP, 1969.

———. *Rapport annuel de l'Institut Français du Pétrole 1969.* IFP, 1970.

———. *Rapport annuel de l'Institut Français du Pétrole 1970.* IFP, 1971.

United Nations. *Convention on the Continental Shelf.* United Nations, Treaty Series, vol. 499 (April 29, 1958).

Jacq, François. "Aux sources de la politique de la science: Mythe ou réalités? (1945– 1970)." *La revue pour l'histoire du CNRS* 6 (2002). https://journals.openedition.org/histoire-cnrs/3611#quotation.

———. "The Emergence of French Research Policy: Methodological and Historiographical Problems (1945–1970)." *History and Technology: An International Journal* 12, no. 4 (1995): 285–308.

Jacquier, Henri. "L'aventure de Jacques-Yves Cousteau, un entretien avec Henri Jacquier," *Mondes francophones*, October 25, 2012. https://mondesfrancophones .com/mondes-europeens/laventure-de-jacques-yves-cousteau-un-entretien-avec -henri-jacquier-3/.

Jamieson, Alan J., Glenn Singleman, Thomas D. Linley, and Susan Casey. "Fear and Loathing of the Deep Ocean: Why Don't People Care about the Deep Sea?" *ICES Journal of Marine Science* 78, no. 3 (July 2021): 797–809.

Jarry, Jean. *L'aventure des bathyscaphes: Marins, ingénieurs et savants au plus profond des mers*. Gerfaut, 2003.

Jasanoff, Sheila, and Kim, Sang-Hyun. "Containing the Atom: Sociotechnical Imaginaries and Nuclear Power in the United States and South Korea." *Minerva* 47, no. 2 (June 2009): 119–46.

———. *Dreamscapes of Modernity. Sociotechnical Imaginaries and the Fabrication of Power*. University of Chicago Press, 2015.

Kaldewey, David, and Désirée Schauz, eds. "Why Do Concepts Matter in Science Policy?" In *Basic and Applied Research: The Language of Science Policy in the Twentieth Century*, edited by David Kaldewey and Désirée Schauz. Berghahn Books: 2018.

Kemp, Alexander G. *The Official History of North Sea Oil and Gas*. Routledge: 2012.

Knott, S. T., and John B. Hersey. "Interpretation of High-Resolution Echo-Sounding Techniques and Their Use in Bathymetry, Marine Geophysics, and Biology." *Deep Sea Research* 4 (1957): 36–44.

Korsmo, Fae L. "The Genesis of the International Geophysical Year." *Physics Today* 60, no. 7 (July 1, 2007): 38–43.

Kreidler, Tai Deckner. *"The Offshore Petroleum Industry: The Formative Years, 1945– 1962."* PhD diss., Texas Tech University, 1997.

Krige, John. *American Hegemony and the Postwar Reconstruction of Science in Europe*. MIT Press, 2008.

Kroll, Gary E. *America's Ocean Wilderness: A Cultural History of Twentieth-Century Exploration*. University Press of Kansas, 2008.

Kullenberg, Börje. *The Piston Core Sampler*. Svenska Hydrografisk-Biologiska Kommissionens Skrifter, ser. 3, vol. 1, no. 2. Elanders Boktryckeri Akitiebolag, 1947. https://epic.awi.de/id/eprint/44416/1/kullenberg_1947.pdf.

Kullenberg, Gunnar. *Ocean Science and International Cooperation: Historical and Personal Recollections*. UNESCO, 2021.

Lacharrière, Guy de la. "Georgette Mariani à la III conférence des Nations Unies sur le droit de la mer." In *Mélanges Juridiques, Rapports Économiques et Juridiques*, 10. Publications du CNEXO, 1981.

Lacombe, Henri. "L'océanographie: Travail d'équipe: Leçon inaugurale du cours d'océanographie physique prononcée le 7 novembre 1955." *Bulletin du Muséum National d'Histoire Naturelle*, 2nd ser., 28, no. 1 (1956): 69–83.

Lacour, Jean-Jacques. *Terre d'innovations 1944–2000: Histoire de l'IFP des origines . . . à nos jours*. Institut Français du Pétrole, 2001.

La Prairie, Yves. *"Ce siècle avait de Gaulle": Un homme de mer témoigne*. Éditions Ouest-France, 1990.

Laubier, Lucien. "L'émergence de l'océanographie au CNRS." *La revue pour l'histoire du CNRS* (online), 6 (2002).

Lawrence, David M. *Upheaval from the Abyss: Ocean Floor Mapping and the Earth Science Revolution.* Rutgers University Press, 2002.

Lelong, Pierre. "Le Général de Gaulle et la recherche en France." *La revue pour l'histoire du CNRS* 1 (1999). http://journals.openedition.org/histoire-cnrs/481.

Le Pichon, Xavier. "Fifty Years of Plate Tectonics: Afterthoughts of a Witness." *Tectonics* 38, no. 8 (August 2019): 2919–33.

———. *Kaiko: Voyage aux extrémités de la mer.* Editions Odile Jacob, 1986.

———. "Mission Nestlante I. 6–23 janvier 1970—Rapport du chef du mission." *Archimer, archive institutionnelle de l'IFREMER.* 1970.

———. "Sea-Floor Spreading and Continental Drift." *Journal of Geophysical Research* 73, no. 12 (1968).

Le Pichon, Xavier, Jean Bonin, Jean Francheteau, and Claude Sibuet. "Une hypothèse d'évolution tectonique du Golfe de Gascogne." In *Histoire structurale du Golfe de Gascogne,* edited by Jacques Debyser, Xavier le Pichon, and Lucien Montadert. Éditions Technip, 1971.

Le Pichon, Xavier, Roy D. Hyndman, and Guy Pautot. "Geophysical Study of the Opening of the Labrador Sea." *Journal of Geophysical Research* 76 (July 1, 1971): 4724–43.

Le Suavé, Raymond, Guy Pautot, Michel Hoffert, Serge Monti, Yann Morel, and Claude Pichocki. "Cadre géologique de concrétions polymétalliques cobaltifères sous-marines dans l'archipel des Tuamotu (Polynésie Française)." *Comptes rendus Académie des Sciences de Paris Série II* 303, no. 11 (1986): 1013–18.

Lidström, Susanna, Sverker Sörlin, and Henrik Svedäng. "Decline and Diversity in Swedish Seas: Environmental Narratives in Marine History, Science and Policy." *Ambio* 49, no. 5 (2022): 1114–21.

Locher, Fabien. "Les pâturages de la guerre froide: Garrett Hardin et la 'tragédie des communs.'" *Revue d'histoire moderne contemporaine* 601, no. 1 (August 19, 2013): 7–36.

———. "Lutter contre l'empire du pétrole." In *Une histoire environnementale de la mer en France et dans l'empire français au XXe siècle.* Mémoire d'habilitation à diriger les recherches en histoire, Science-Po Paris, 2022 (unpublished).

Lofi, Johanna, Christian Gorini, Serge Berné, et al. "Erosional Processes and Paleo-Environmental Changes in the Western Gulf of Lions." *Marine Geology* 217, no. 1–2 (May 2005): 1–30.

"Loi n° 68–1181 du 30 décembre 1968 relative à l'exploitation du plateau continental et à l'exploitation de ses ressources naturelles." *Journal officiel de la République Française* 307 (December 1968).

Lore, G. L. "An Exploration and Discovery Model: A Historic Perspective—Gulf of Mexico Outer Continental Shelf." Proceedings of the International Conference on Arctic Margins, Magadan, Russia, 1994. https://www.boem.gov/about-boem/icam -1994-conference-proceedings-magadan-russia.

Madsen, Axel. *Cousteau: An Unauthorized Biography.* Beaufort Books, 1986.

Magny, Jules. *L'océan.* A. Rigaud, 1876.

Malod, Jacques-André, and Jean Mascle. "Structures géologiques de la marge continentale à l'ouest du Spitzberg." *Marine Geophysical Researches* 2 (1974): 215–29.

Mann Borgese, Elisabeth. *The Mines of Neptune.* H. N. Abrams, 1985.

Martínez-Rius, Beatriz. "For the Benefit of All Men: Oceanography and Franco-American Scientific Diplomacy in the Cold War, 1958–1970." *Berichte zur Wissenschaftsgeschichte* 43, no. 4 (November 2020): 581–605.

————. "An Open Secret: Marine Geosciences, Offshore Oil Exploration and Industrial Secrecy in the Mediterranean Seafloor." *History and Technology* 40, no. 3 (2024): 173–200.

Mascle, Jean, ed. *Des laboratoires de zoologie marine à l'Observatoire Océanologique de Villefranche-sur-Mer.* Observatoire Océanologique Villefranche-sur-Mer, 2010.

Matsen, Bradford. *Jacques Cousteau: The Sea King.* Pantheon Books, 2009.

McDougall, Walter A. "Space-Age Europe: Gaullism, Euro-Gaullism, and the American Dilemma." *Technology and Culture* 26, no. 2 (April 1985): 179–203.

McNeil, Maureen, Michael Arribas-Ayllon, Joan Haran, Adrian Mackenzie, and Richard Tutton. "Conceptualizing Imaginaries of Science, Technology, and Society." In *The Handbook of Science and Technology Studies*, 4th ed., edited by Ulrike Felt. MIT Press, 2017.

McQuillin, Robert. *An Introduction to Seismic Interpretation: Reflection Seismics in Petroleum Exploration.* Gulf, 1984.

Médioni, René. "L'oeuvre des géologues français au Maroc." *Travaux du Comité Français d'Histoire de la Géologie, ser.*3, vol. 25, no. 1 (2011): 1–52.

Menard, William H., Stuart M. Smith, and Richard M. Pratt. "The Rhône Deep-Sea Fan." In *Submarine Geology and Geophysics*, edited by Walter F. Whittard and R. Bradshaw. Colston Papers, no. 17. Butterworths, 1965.

Mero, John. *The Mineral Resources of the Sea.* Elsevier, 1965.

Merrell, William J., Mary Hope Katsouros, and Jacqueline Bienski. "The Stratton Commission: The Model for a Sea Change in National Marine Policy." *Oceanography* 14, no. 2 (2001): 11–16.

Michelet, Jules. *La mer. 5th ed.* Michele Lévy Frères, 1875.

Miles, Edward L. *Global Ocean Politics: The Decision Process at the Third United Nations Conference on the Law of the Sea 1973—1982.* Nijhoff, 1998.

————. "Technology, Ocean Management, and the Law of the Sea." *Denver Law Journal* 46 (1969): 240–60.

Montadert, Lucien, Jean Sancho, Jean-Pierre Fail, and Jacques Debyser. "De l'âge tertiaire de la série salifère responsable des structures diapiriques en Méditerranée occidentale (nord-est des Baléares)." *Comptes rendus de l'Académie des Sciences* 271 (1970): 812–15.

Morgan, W. Jason. "Rises, Trenches, Great Faults, and Crustal Blocks." *Journal of Geophysical Research* 73, no. 6 (1968): 1959–82.

Morvan, R. G., ed. *Les Océans.* Morvan humanisme: Série les grands enquêtes no. 24–25 (1971).

Mounecif, Radouan Andrea. *"Chercheurs d'or noir: Les pétroliers français entre le Sahara et le monde (1924–2003)".* PhD diss., Sorbonne Université, 2021.

Mukerji, Chandra. *A Fragile Power: Scientists and the State.* Princeton University Press, 1989.

Muraour, Pierre, Olivier Leenhardt, and Jacques Merle. "Eléments pour un programme de recherches séismiques en Méditerranée." In *Océanographie géologique et géophysique de la Méditerranée occidentale: Villefranche sur Mer, 4 au 8 Avril 1961*, edited by Colloques nationaux du CNRS. Éditions du Centre National de la Recherche Scientifique, 1962.

Muraour, Pierre, Jacques Merle, and Jean Ducrot. "Observations sur le plateau continental à la suite d'une étude séismique par réfraction dans le Golfe du Lion." *Académie des Sciences de Paris* 254, no. 15 (April 9, 1962): 2801–3.

National Academies of Sciences, Engineering, and Medicine. *High-Performance Bolting Technology for Offshore Oil and Natural Gas Operations.* National Academies Press, 2018.

———. "History of the Offshore Oil and Gas Industry and the Development of Safety Efforts." In *Strengthening the Safety Culture of the Offshore Oil and Gas Industry.* National Academies Press, 2016.

Nesteroff, Wladimir. "La crise de salinité méssinienne en Méditerranée: Enseignements des forages JOIDES et du bassin de Sicile." In *Rapports et procès-verbaux des réunions* 22, edited by Commission Internationale pour l'Exploration Scientifique de la Mer Méditerranée (CIESM). CIESM, 1973.

Noble Shor, Elizabeth. *Scripps Institution of Oceanography: Probing the Oceans, 1936 to 1976.* Tofua Press, 1978.

Nouschi, André. *La France et le pétrole: De 1924 à nos jours.* Picard, 2001.

———. *Pétrole et relations internationales depuis 1945.* Armand Collin, 1999.

Nyaberg, Roberto. "A Few Strategic Considerations Concerning French Imports of Oil Products Between 1946 and 2005." In *Oil Producing Countries and Oil Companies,* edited by Alain Beltran. Peter Lang, 2011.

Olivet, Jean-Louis, Bertrand Sichler, Pierre Thonon, Xavier le Pichon, Guy Martinais, and Guy Pautot. "La faille transformante Gibbs entre le rift et la marge du Labrador." *Comptes Rendus Académie des Sciences de Paris* 271 (September 1, 1970): 949–52.

Oreskes, Naomi. "A Context of Motivation: US Navy Oceanographic Research and the Discovery of Sea-Floor Hydrothermal Vents." *Social Studies of Science* 33, no. 5 (2003): 697–742.

———. "The Iron Curtain of Classification: What Difference Did It Make?" In *Science on a Mission: How Military Funding Shaped What We Do and Don't Know about the Ocean.* University of Chicago Press, 2021.

———. *Science on a Mission: How Military Funding Shaped What We Do and Don't Know about the Ocean.* University of Chicago Press, 2021.

Oreskes, Naomi, and H. E. LeGrand, eds. *Plate Tectonics: An Insider's History of the Modern Theory of the Earth.* Westview Press, 2003.

Paccalet, Yves. *Jacques-Yves Cousteau. Dans l'océan de la vie.* JC Lattès, 1997.

Pautot, Guy. "La dorsale medio-atlantique et le renouvellement des fonds océaniques." *Revue de géographie physique et de géologie dynamique* 2, vol. 6, no. 5 (1970): 379–402.

———, ed. *Résultats de la campagne de flexo-électro-carottage en Méditerranée nord-occidentale.* Publications du Centre National pour l'Exploitation des Océans, January 1972.

———. "TRANSPAC rapport de mission." CNEXO (1973).

Pautot, Guy, Jean-Marie Auzende, and Xavier le Pichon. "Continuous Deep Sea Salt Layer along North Atlantic Margins Related to Early Phase of Rifting." *Nature* 227, no. 5256 (July 1970): 351–54.

Pestre, Dominique. "L'évolution des champs de savoir, interdisciplinarité et noyaux durs communication au conseil scientifique du CNRS (14 janvier 2002)." Natures Sciences Sociétés 12, no. 2 (2004): 191–96.

Peters, Kimberly, Philip Steinberg, and Elaine Stratford. *Territory Beyond Terra.* Rowman and Littlefield, 2018..

Peterson, Melvin N. A., et al. *Initial Reports of the Deep Sea Drilling Project. Vol. 2.* US Government Printing Office, 1970.

Pettersson, Hans. *Westward Ho with the Albatross*. E. P. Dutton, 1953.

Philippe, Sébastien, and Tomas Statius. *Toxique: Enquête sur les essais nucléaires Français en Polynésie*. Presses Universitaires de France/Humensis, 2021.

Piketty, Gérard. "Allocution de Gérard Piketty, président du CEP&M, à la journée CEP&M-COPREP du 10 Octobre 2000, marquant par ailleurs le 50ème anniversaire du FSH." In *Conference Journées du pétrole CEP&M-COPREP*, Paris, October 10–11, 2000. Comité d'Études Petrolières et Marines, 2002.

Mysteries of the Deep: How Seafloor Drilling Expeditions Revolutionized Our Understanding of Earth History. MIT Press, 2024.

Pratt, Joseph A., Tyler Priest, and Christopher James Castaneda. *Offshore Pioneers: Brown and Root and the History of Offshore Oil and Gas*. Gulf, 1997.

Priest, Tyler. "Extraction Not Creation: The History of Offshore Petroleum in the Gulf of Mexico." *Entreprise and Society* 8, no. 2 (June 2007): 227–67.

———. *The Offshore Imperative: Shell Oil's Search for Petroleum in Postwar America*. Texas A&M University Press, 2010.

Pritchard, Sara B. *Confluences: The Nature of Technology and the Remaking of the Rhône*. Harvard University Press, 2011.

Prodehl, Claus, and Walter D. Mooney. *Exploring the Earth's Crust: History and Results of Controlled-Source Seismology*. Geological Society of America, 2012.

Quéneudec, Jean Pierre. "La France et le droit de la mer." In *The Law of the Sea*, ed. T. Treves. Kluwer Law International, 1997.

Rainger, Ronald. "Science at the Crossroads: The Navy, Bikini Atoll, and American Oceanography in the 1940s." *Historical Studies in the Physical and Biological Sciences* 30, no. 2 (2000): 349–371.

Ranganathan, Surabhi. "Decolonization and International Law: Putting the Ocean on the Map." *Journal of the History of International Law/Revue d'Histoire du Droit International* 23, no. 1 (December 10, 2020): 161–83.

———. "Global Commons." *European Journal of International Law* 27, no. 3 (August 2016): 693–717.

———. "Ocean Floor Grab: International Law and the Making of an Extractive Imaginary." *European Journal of International Law* 30, no. 2 (July 22, 2019): 573–600.

Reidy, Michael S., and Helen M. Rozwadowski. "The Spaces In Between: Science, Ocean, Empire." *Isis* 105, no. 2 (2014): 338–51.

Riffaud, Claude, and Xavier le Pichon. *Expédition «FAMOUS»: À 3000 mètres sous l'Atlantique*. Albin Michel, 1976.

Robelius, Fredrik. *"Giant Oil Fields: The Highway to Oil, Giant Oil Fields, and Their Importance for Future Oil Production."* PhD diss., Faculty of Science and Technology of Uppsala University, 2007.

Roberts, Peder. "Intelligence and Internationalism: The Cold War Career of Anton Bruun." *Centaurus* 55, no. 3 (August 2013): 243–63.

Robinson, Sam. "Early Twentieth-Century Ocean Science Diplomacy: Competition and Cooperation among North Sea Nations." *Historical Studies in the Natural Sciences* 50, no. 4 (September 23, 2020): 384–410.

———. *Ocean Science and the British Cold War State*. Palgrave Macmillan, 2018.

———. "Scientific Imaginaries and Science Diplomacy: The Case of Ocean Exploitation." *Centaurus* 63, no. 1 (2021): 150–70.

Rosaire, Eugene E., and Oliver C. Lester. "Seismological Discovery and Partial Detail of Vermilion Bay Salt Dome, Louisiana." *AAPG Bulletin* 16 (1932): 1221–29.

Ross, Donald. "Twenty Years of Research at the SACLANT ASW Research Center, 1959–1979." NATO, 1980.

Roveri, Marco, Vinicio Manzi, A. Bergamasco, et al. "Dense Shelf Water Cascading and Messinian Canyons." *American Journal of Science* 314, no. 3 (March 2014): 751–84.

Rozwadowski, Helen M. "Arthur C. Clarke and the Limitations of the Ocean as a Frontier." *Environmental History* 17, no. 3 (2012): 578–602.

———. "Bringing Humanity Full Circle Back into the Sea." *Environmental Humanities* 14, no. 1 (March 1, 2022): 1–28.

———. "Engineering, Imagination, and Industry: Scripps Island and Dreams for Ocean Science in 1960." In *The Machine in Neptune's Garden: Historical Perspectives on Technology and the Marine Environment*, edited by Helen M. Rozwadowski and David K. Van Keuren. Science History Publications, 2004.

———. "Ocean's Depths." *Environmental History* 15, no. 3 (July 1, 2010): 520–25.

———. "Oceans: Fusing the History of Science and Technology with Environmental History." In *A Companion to American Environmental History*, edited by Douglas Cazaux Sackman. Wiley-Blackwell, 2010.

———. "The Promise of Ocean History for Environmental History." *Journal of American History* 100, no. 1 (June 1, 2013): 136–39.

———. *The Sea Knows No Boundaries: A Century of Marine Science under ICES*. International Council for the Exploration of the Sea, University of Washington Press, 2002.

———. *Vast Expanses: A History of the Oceans*. Reaktion Books, 2019.

Rozwadowski, Helen M., and David K. Van Keuren, eds. *The Machine in Neptune's Garden: Historical Perspectives on Technology and the Marine Environment*. Science History Publications, 2004.

Ruffini, Pierre-Bruno. "The Intergovernmental Panel on Climate Change and the Science-Diplomacy Nexus." *Global Policy* 9 (November 2018): 73–77.

Rusnak, Gene A. "Afoot and Afloat Along the Edge: Adventures of an Ingenious Beachcomber—A Tribute to Francis Parker Shepard (1897–1985)." In *From Shore to Abyss: Contributions in Marine Geology in Honor of Francis Parker Shepard*, edited by Robert H. Osborne. Society for Sedimentary Geology, 1991.

Ryan, William B. F. *"The Floor of the Mediterranean Sea."* PhD diss., Columbia University, 1969.

Ryan, William B. F., Kenneth J. Hsü, and Maria B. Cita. "The Origin of the Mediterranean Evaporites." In *Initial Reports of the Deep Sea Drilling Project*, vol. 13, edited by William B. F. Ryan, Kenneth J. Hsü, et al. US Government Printing Office, 1973.

Ryan, William B. F., Kenneth J. Hsü, et al. *Initial Reports of the Deep Sea Drilling Project*. Vol. 13. US Government Printing Office, 1973.

Sabol Spezio, Teresa. *Slick Policy: Environment and Science Policy in the Aftermath of the Santa Barbara Oil Spill*. University of Pittsburgh Press, 2018.

Salut, Samir. "Politique nationale du pétrole, sociétés nationales et 'Pétrole Franc.'" *Revue historique* 2, no. 638 (2006): 355–88.

Sapolsky, Harvey M. *Science and the Navy: The History of the Office of Naval Research*. Princeton University Press, 1990.

Sayle, Timothy Andrews. *Enduring Alliance: A History of NATO and the Postwar Global Order*. Cornell University Press, 2019.

Scharf, Michael P. "The Truman Proclamation on the Continental Shelf." In *Customary International Law in Times of Fundamental Change: Recognizing Grotian Moments.* Cambridge University Press, 2013.

Schiavon, Martina. *Itinéraires de la précision: Géodésiens, artilleurs, savants et fabricants d'instruments de précision en France, 1870–1930.* Presses Universitaires de Lorraine, 2014.

Schiefelbein, Susan, and Jacques Cousteau. *The Human, the Orchid, and the Octopus: Exploring and Conserving Our Natural World.* Bloomsbury USA, 2007.

Schlich, Roland. *BENIN 1971 Cruise, RV Jean Charcot.* Archimer, archive institutionnelle de l'IFREMER.

Schwach, Vera. "The Sea Around Norway: Science, Resource Management, and Environmental Concerns, 1860–1970," *Environmental History* 18, no. 1 (2013): 101–10.

Selli, Raimondo. "Il bacino del metauro: Descrizione geologica, risorse mineraire, idrogeologia." *Giornale di geologia* 2, no. 24 (1952): 1–80.

———. "An Outline of the Italian Messinian." In *Messinian Events in the Mediterranean,* edited by C. M. Drooger. North Holland, 1973.

Sénat. "Rapport d'information—L'exploration, la protection et l'exploitation des fonds marins: Quelle stratégie pour la France?" Senate ordinary session, 2021–2022, no. 723. June 21, 2022.

Sewell, Seymour R. B., and John D. H. Wiseman. "Le relief du fond de la mer, particulièrement dans l'hémisphère méridional." In *Comptes rendus du Congrès International de Géographie,* edited by the International Geographic Congress. E. J. Brill, 1938.

Shepard, Francis P. "American Submarine Canyons." *Scottish Geographical Magazine* 50, no. 4 (July 1, 1934): 212–18.

———. *Submarine Geology.* Harper, 1948.

———. "The Underlying Causes of Submarine Canyons." *Proceedings of the National Academy of Sciences of the United States of America* 22, no. 8 (August 1936): 496–502.

Sigsbee, Charles D. *Deep-Sea Sounding and Dredging: A Description and Discussion of the Methods and Appliances Used on Board the Coast and Geodetic Survey Steamer "Blake."* US Government Printing Office, 1880.

Simoncini, Nicolas. "*Histoire de la recherche sur les piles à combustible en France des années soixante aux années quatre-vingt.*" PhD diss., Université de Technologie de Belfort-Montbéliard, 2018.

Sismondo, Sergio. "Sociotechnical Imaginaries: An Accidental Themed Issue." *Social Studies of Science* 50, no. 4 (August 1, 2020): 505–7.

Smith, Jason W. *To Master the Boundless Sea: The U.S. Navy, the Marine Environment, and the Cartography of Empire.* University of North Carolina Press, 2018.

Sörlin, Sverker, and Nina Wormbs. "Environing Technologies: A Theory of Making Environment." *History and Technology* 34, no. 2 (April 3, 2018): 101–25.

Squire, Rachael. *Undersea Geopolitics: Sealab, Science, and the Cold War.* Rowman and Littlefield, 2021.

Starosielski, Nicole. *The Undersea Network.* Duke University Press, 2015.

Starr, Chauncey. "Andre Y. Giraud 1925–1997." *Memorial Tributes: National Academy of Engineering,* vol. 10, 107–9.

Steinbeck, John. *The Log from the Sea of Cortez.* Penguin Books, 1951.

Steinberg, Philip E. *The Social Construction of the Ocean*. Cambridge University Press, 2001.

Sternlicht, Daniel D. "Looking Back: Seismic Exploration." *IEEE Potentials* 18, no. 1 (February–March 1999): 36–38.

Suarez, Suzette V. *The Outer Limits of the Continental Shelf: Legal Aspects of Their Establishment*. Springer-Verlag, 2008.

Sykes, Lynn R. "Mechanism of Earthquakes and Nature of Faulting on the Mid-Oceanic Ridges." *Journal of Geophysical Research (1896–1977)* 72, no. 8 (1967): 2131–53.

Thompson, Thomas L. "Plate Tectonics in Oil and Gas Exploration of Continental Margins." *AAPG Bulletin* 60, no. 9 (September 1, 1976): 1463–1501.

Turchetti, Simone. *Greening the Alliance: The Diplomacy of NATO's Science and Environmental Initiatives*. University of Chicago Press, 2019.

———. "Sword, Shield and Buoys: A History of the NATO Sub-Committee on Oceanographic Research, 1959–1973." *Centaurus* 54, no. 3 (2012): 205–31.

Turchetti, Simone, and Peder Roberts, eds. *The Surveillance Imperative*. Palgrave Macmillan US, 2014.

Turner, Frederik Jackson. *The Significance of the Frontier in American History*. 1893.

UNESCO-IOC. *The United Nations Decade of Ocean Science for Sustainable Development (2021–2030): Implementation Plan*. IOC Ocean Decade Series, vol. 19. UNESCO, 2021.

United Nations. *Convention on the High Seas 1958*. United Nations Treaty Series, vol. 450, April 29, 1958.

———. *Convention on the Law of the Sea*. United Nations Treaty Series, vol. 1833, December 10, 1982.

———. *Third Meeting, 3 March 1958*. Official Records of the United Nations Conference on the Law of the Sea, vol. 6 (Fourth Committee: Continental Shelf). United Nations, 1958.

United Nations General Assembly. First Committee Debate, UN Doc. A/PV.1515 para. 91. November 1, 1967. UN Digital Library.

———. *Declaration of Principles Governing the Sea-Bed and the Ocean Floor, and the Subsoil Thereof, beyond the Limits of National Jurisdiction*. A/RES/2749(XXV). November 17, 1970. UN Digital Library.

———. "Note Verbal from the Permanent Mission of Malta to the United Nations Addressed to the Secretary-General," UN Doc. A/6695. August 18, 1967. UN Digital Library.

———. Resolution 2749 (XXV) of December 17, 1970, 25th Session. UN Digital Library.

United States Patent and Trademark Office. *Apparatus for Firing Explosive Charges Under Water*. US patent 3,360,070. Filed December 26, 1967.

Vanderpool, Christopher K. "Marine Science and the Law of the Sea." *Social Studies of Science* 13, no. 1 (February 1983): 107–29.

Van Keuren, David K. "Breaking New Ground: The Origins of Scientific Ocean Drilling." In *The Machine in Neptune's Garden: Historical Perspectives on Technology and the Marine Environment*, edited by Helen M. Rozwadowski and David K. Van Keuren. Science History Publications, 2004.

Vanney, Jean-René, and Jean-Pierre Pinot. "La campagne du *Discoverer* au large de l'Afrique tropicale." *Bulletin de l'Association de Géographes Français*, année 1969: 368–69.

Vindt, Gérard. "De la CFP à Total, une entreprise sous contrôle de l'Etat." *Alternatives Economiques* 295, no. 10 (October 28, 2010): 111.

Vine, Frederick, and Drummond Matthews. "*Magnetic Anomalies over Oceanic Ridges*." *Nature* 199 (1963): 947–49.

Waldock, C. H. M. "The Legal Basis of Claims to the Continental Shelf." *Transactions of the Grotius Society* 36 (1950): 115–48.

Watts, Jonathan. "Race to the Bottom: The Disastrous, Blindfolded Rush to Mine the Deep Sea." *The Guardian*, September 27, 2021.

Weir, Gary E. *An Ocean in Common: American Naval Officers, Scientists, and the Ocean Environment*. Texas A&M University Press, 2001.

Wenk, Edward Jr. *Making Waves: Engineering, Politics, and the Social Management of Technology*. University of Illinois Press, 1995.

———. *The Politics of the Ocean*. University of Washington Press, 1972.

Wertenbaker, William. *The Floor of the Sea*. Little Brown, 1974.

White, Richard, Patricia Nelson Limerick, and James R. Grossman. *The Frontier in American Culture: An Exhibition at the Newberry Library, August 26, 1994—January 7, 1995*. University of California Press, 1994.

Whittard, Walter Frederick, and R. Bradshaw, eds. *Submarine Geology and Geophysics. Colston Papers, no. 17*. Butterworths, 1965.

Wilson, J. Tuzo. "A New Class of Faults and Their Bearing on Continental Drift." *Nature* 207 (July 1965): 343–47.

Winthrop Haight, George. "United Nations Affairs: Ad Hoc Committee on Sea-Bed and Ocean Floor." *International Lawyer* 3, no. 1 (1969): 22–30.

Wrigley, Charlotte. "Going Deep: Excavation, Collaboration and Imagination at the Kola Superdeep Borehole." *Environment and Planning D: Society and Space* 41, no. 3 (2023): 549–67.

Yergin, Daniel. *The Prize: The Epic Quest for Oil, Money and Power*. Free Press, 2009.

Young, Richard. "The Legal Regime of the Deep-Sea Floor." *American Journal of International Law* 62, no. 3 (1968): 641–53.

Zelko, Frank. *Make It a Green Peace! The Rise of a Countercultural Environmentalism*. Oxford University Press, 2013.

Index

academic-industrial collaboration. *See* industrial-academic collaboration

AFERNOD (French Association for the Study and Research on Oceanic Nodules), 135, 137

Albania: geology and geography of, 22; in World War I, 20–22

Algeria: coup d'état, 54–55; Evian Agreements and oil supplies, 76; as French colonial territory, 22–23, 133; French oil industry and, 23, 35, 75–76; independence of, 5, 16, 55, 75–76, 137. *See also* de Gaulle, Charles

Alvin (submersible), 124–26. *See also* FAMOUS (Franco-American Mid-Ocean Undersea Study); submersibles

Alwin North (oil and gas field), 164

American Petroleum Institute (API), 33, 34

Amoco Cadiz oil spill, 168

anticipation to resource production at sea, 6, 32, 53–54, 77–78, 109–10, 118–19, 130–31, 169–71. *See also* Mediterranean: hydrocarbon potential; Truman Proclamation

ANZIC (Australia and New Zealand Consortium), 178

applied ocean research: and France, 51, 57; and imbalance with basic research, 51–52; and marine geosciences, 87–88, 108. *See also* CNEXO (National Center for the Exploitation of the Oceans); fisheries: and research

Arab-Israeli war, 75

Archimède (submersible), 124–26, 128. *See also* FAMOUS (Franco-American Mid-Ocean Undersea Study); submersibles

Atlantic Ocean (North): industrial-academic collaboration, 83–84, 100–101, 105–8; industrial surveys, 81, 83; international relations, 17, 97, 100, 113–14, 121–22, 124–26; military surveillance, 49–50, 55, 57; seafloor research, 16–17, 25–27, 38, 62, 65, 83–84, 97–101, 106–8, 125–26, 151, 158; territory of France, 5, 54, 116

Autonomous Board of Petroleum, 77, 79

Auzende, Jean-Marie, 134

Balanceanu, Jean-Claude, 97, 100

Balearic basin, 71–72, 105, 155–56, 161. *See also* Mediterranean: salt domes

basic ocean research: in France's science policy, 51, 56–57, 59, 87, 97, 173; in international cooperation, 48, 51, 129. *See also* applied ocean research: and imbalance with basic research

bathymetry, 24–25

Bay of Biscay, 26, 82, 100; geological history, 106–8; hydrocarbon potential, 82, 107

Behm, Alexander, 25. *See also* echo-sounders

benefit of humankind. *See* "common heritage of humankind"

big science, marine geosciences as, 111, 141. *See also* Deep Sea Drilling Project (DSDP)

ENIEPSA, 163
Entreprise de Recherches et d'Activités
  Pétrolières (ERAP), 80, 84, 95
environmental awareness of oceans, 14,
  165–66, 168
environmental impact: of geophysical
  surveys, 80–81; of oil spills, 168
Ewing, Maurice: geophysics and, 22,
  60–65, 98; relationship with le Pichon,
  98; research on age of ocean crust, 65;
  research on Gulf of Mexico, 72, 147,
  150; research on Mediterranean, 43,
  151; in scientific ocean drilling, 147–49
exploitation, definition of, 11
Eyrès, Marc-Marie, 58

Fail, Jean-Pierre, 80
FAMOUS (Franco-American Mid-Ocean
  Undersea Study), 17, 113, 114,
  123–29; mineral resources and, 129
Fessenden, Reginald, 25. *See also*
  echo-sounders: development of
fisheries: commercial, 32, 52, 167; control
  on high seas, 4, 52; and research, 47–
  48, 51, 57, 58, 132, 134. *See also* applied
  ocean research
Flexotir: academic-industrial surveys,
  96–97, 99–102, 106 (*see also*
  continental shelf: mapping by
  CNEXO; Guinée I [expedition];
  Jean Charcot [research vessel];
  Nestlante [expeditions]; Noratlante
  I [expedition]); capabilities, 80–81;
  industrial surveys, 81
France: colonial empire, 5, 22–23, 57,
  75–77, 110; Fifth Republic, 55–56 (*see
  also* de Gaulle, Charles: presidency);
  maritime territory, 5–6, 51; oil industry
  (*see* oil industry [France]); overseas
  territories, 5–6, 22, 51, 54 (*see also*
  French Polynesia); policy on science
  and technology (*see* basic ocean
  research: in France's science policy; de
  Gaulle, Charles: presidency; General
  Delegation for Scientific and Technical
  Research [DGRST])
France-Japan relations through ocean
  exploration, 114, 118, 127–28, 138, 178

France-US relations, 55, 119–20; bilateral
  agreement on ocean exploration,
  122–23; comparison of research
  capabilities, 60–61, 65, 72, 85–86, 118–
  19, 122; through ocean exploration,
  119–29 (*see also* FAMOUS [Franco-
  American Mid-Ocean Undersea
  Study]). *See also* de Gaulle, Charles:
  presidency; NATO: France's position
  in
French Association for the Study and
  Research on Oceanic Nodules
  (AFERNOD), 135, 137
French Institute of Petroleum (IFP):
  academic collaborations, 95–
  97, 99–101, 106–8, 159, 160–62;
  industrial institutions and, 75,
  77–78, 163; Marine Project, 79–82;
  Mediterranean Project, 159–60;
  offshore development and, 79–83,
  105, 106, 160–61, 163 (*see also*
  Committee of Petroleum and Marine
  Studies [CEPM]; *Flexotir*)
French Navy, 6, 12, 29, 35, 36, 37, 38,
  43, 70, 86, 89, 116, 134. *See also*
  Hydrographic Service
French Polynesia, 5, 10, 114, 130, 133–34,
  137, 151–53
French Research Institute for Exploitation
  of the Sea (IFREMER), 128, 175
Frigg (gas field), 164
Fuels Directorate, 76, 78–79
Furnestin, Jean, 58

Gauthier, Michel, 132
General Delegation for Scientific and
  Technical Research (DGRST):
  France's science policy and, 56;
  support to ocean research, 56–60, 70,
  74, 85–86, 88, 124
Gennesseaux, Maurice, 36
Geographical Service of the French
  Army, 21
Geological and Mining Research
  Office (BRGM), 106–7, 110. *See
  also* continental shelf: mapping by
  CNEXO
Geological Society of America, 62

168; military surveillance in (*see* French Navy; Hydrographic Service; NATO: in Mediterranean); salt domes, 71–72, 105, 140–42, 148–50, 154; sea-level regression, 20, 22–24, 29, 38, 40, 72, 141, 164 (*see also* Messinian Salinity Crisis; salt domes). *See also* Deep Sea Drilling Project (DSDP): Mediterranean expeditions

Mediterranean salt giant.
    See Mediterranean: salt domes; Mediterranean: sea-level regression
Menard, Henry W., 72, 150, 154
Mero, John L., 130, 135
Messinian Salinity Crisis, 157–61, 164
Metal Gesellschaft AG, 130
military patronage: in France's ocean sciences, 12, 15 (*see also* Bourcart, Jacques: military patronage of); in US ocean sciences, 4, 8, 173–74
Ministry of Industry (France), 11, 16, 51, 76, 78, 79, 158, 159
Mitsubishi Heavy Industries and deep-sea mining, 131
Mitterrand, François, 175
Mohole Project, 144. See also *CUSS I* (drillship)
Monde, Le, publications on ocean sciences, 80, 93, 95, 99, 140, 148, 158, 167, 168
Montadert, Lucien, 108, 160–62, 164
Morgan, Jason, 96
Morocco, 21, 23, 35, 100, 101
Muraour, Pierre, 106
Museum of Monaco. *See* Oceanographic Museum of Monaco

National Commission on Marine Science, Engineering, and Resources, 119
National Geographic Society, 65–66, 125
NATO: France's position in, 12, 55, 119; in Mediterranean, 51, 55; Subcommittee on Oceanographic Research (ORC), 50; support to ocean research, 50–51, 56, 58
Nautile (submersible), 127. *See also* KAIKO (expedition)
Navarre, André, 108
neocolonialism in the oceans, 77, 110, 115

NERC (Natural Environment Research Council), 109
Nesteroff, Wladimir, 153
Nestlante (expeditions), 16, 103–10
"new deep territories," definition of, 174–75
Noratlante I (expedition), 16, 97, 99–100, 108
North Sea. *See* offshore oil industry: in North Sea
Norway, 1, 77, 102–3. *See also* offshore oil industry: in North Sea
NSF (National Science Foundation), 18, 141, 144, 145, 148, 153, 178
nuclear tests (France). *See* Pacific Ocean: nuclear tests in

ocean diplomacy, 17, 109–10, 113–15, 116–20, 134, 137–38, 152–53, 171–72, 174, 177. *See also* France-Japan relations through ocean exploration; France-US relations: through ocean exploration
ocean governance: *res communis*, 31; *res nullius*, 31, 115; UNCLOS I, 52, 53–54, 94–95, 153; UNCLOS III, 117
ocean research: compared to nuclear research, 13, 90–91; compared to space research, 13, 56, 88, 90–91, 124, 144; for international prestige, 43–44, 47–49, 50, 70, 85, 98, 114, 118–19, 121, 125, 129, 167; and resource exploration (*see* industrial-academic collaboration); for territorial control, 57, 85, 109–10, 113, 137
Oceanographic Museum of Monaco, 42, 65–66, 84. *See also* Cousteau, Jacques-Yves
Oceanological Center of Brittany (COB), 114, 121–22, 136; modeled on American research centers, 122
oceans: conquest of (rhetoric), 3, 77, 87, 92, 113, 169, 175, 177; and diplomacy (*see* ocean diplomacy); popular perception of, 2, 6, 14, 168, 175–76; as territories, 77–78 (*see also* seafloor: as territory)
Office of Naval Research, 62

INDEX › 239